Fascination Quadrocopt

Die Deutsche Nationalbibliothek verzeichnet diese Publikation in der Deutschen Nationalbibliografie; detaillierte bibliografische Daten sind im Internet über dnb.d-nb.de abrufbar.

Impressum:

© 2011 Roland Büchi
Herstellung und Verlag: Books on Demand GmbH, Norderstedt

ISBN: 978-3-8423-6731-9

Die Deutsche Fassung ist unter dem Titel: 'Faszination Quadrokopter' beim vth-Verlag erschienen.

Preamble

For Quadrocopters there are many names. They are also called Quadrotor, Quadricopter, Microcopter, Multicopter or simply 'UFO'. This type of model construction is very young. It began early in the new millennium with commercial products for flying camera inspections. In 2005, the company Silverlit sold an inexpensive toy with four horizontally arranged propellers. So the hobbyists were able to have their first experience with the new model aircraft.

It was the current microprocessor technology which opened up this fascinating subject of model construction. It is able to evaluate sensors for attitude stabilization and change the speeds of the motors immediately. This book explains the working principle, the used motors, sensors and control systems.

For the construction of a Quadrocopter, components are often only offered by different manufacturers. The model builder must execute the assembly and the putting into operation himself. Therefore this book also contains tips for the wiring and the making or purchase of the frame. This needs to be lightweight and stable. It also carries the components.

The chapters 'Flight mechanics', 'Setting the controller' and 'Dimensioning of motors and propellers' include a bit more theory. This is only meant as a link to the practice and will run only as far as is necessary to understand the basic functioning. The main findings are always shown in brief at the end of these chapters, based on the theory.

The setting of the regulator is of great importance. In many control systems, it is possible to download new software updates and individual parameters via a PC interface. Thus, in each Quadrocopter, a different behavior can be achieved according to taste. The aerobatic pilot wants an agile behavior. The novice flyer or photo flyer prefers more of a good-natured behavior.

The autonomous navigation using GPS and creation of photos and movies from the air today enjoys growing popularity as a hobby and also for commercial use. These Quadrocopters serving as camera platforms also find their place in this book.

It is important to take note that there are special rules for different countries on the use of Quadrocopters. Especially the use of GPS and the autonomous flight often requires special permits. Each Quadrocopter pilot should therefore check on the laws with the state government.
References to literature are listed at the end of the book.

Contents

1. **Functionality** 7
   1.1 Steering mechanism 7
   1.2 Physical movement 8
   1.3 Flight in '+' or 'x' configuration 10

2. **Base components** 13
   2.1 Control board 14
   2.2 Sensors 15
   2.3 Brushless motors, propellers 20
   2.4 Brushless controllers 23
   2.5 Lithium-polymer accumulator 25
   2.6 Remote control and receiver 28
   2.7 PC interface 30
   2.8 Frame construction 31
   2.9 Cabling 36
   2.10 Safety 38

3. **Extension components** 40
   3.1 Air pressure sensor 42
   3.2 Compass 44
   3.3 GPS 46
   3.4 Onboard cameras, video-recording 49

4. **Flight mechanics** 52
   4.1 Hover flight 52
   4.2 The attack angle φ 53
   4.3 General balance of forces 58
   4.4 A simple physical model 60
   4.5 Findings in brief 63

5. **Setting the controller** 64
   5.1 Control of nick and roll axis 64

    5.2   Effect of K_P and K_D 66
    5.3   Transfer function 71
    5.4   Heading hold 73
    5.5   Findings in brief 75

6.   Dimensioning of motors and propellers 77
    6.1   Propellers 77
    6.2   Larger propellers with gearbox 81
    6.3   Motor 83
    6.4   Three- or four-blade propellers 86
    6.5   Power and thrust measurement 86
    6.5   Findings in brief 88

7.   Special shapes, Tri-, Hexa-, Octocopters 89
    7.1   Tricopters 89
    7.2   Hexacopters, Octocopters 94
    7.3   Depron bodies 96
    7.4   Aerial Sedan 100
    7.5   Private transport with Quadrocopters from today's perspective 105

8.   Initial operation, sources of error and first flight 108
    8.1   Check the functions on the ground 108
    8.2   Range test of the remote control 109
    8.3   Mounting the propellers 110
    8.4   First flight 112

9.   Literature 115

1. Functionality

1.1 Steering mechanism

Quadrocopters are aircraft with four propellers. They have the same control capabilities as helicopters. Figure 1 illustrates this. The stick assignment of the remote control, as shown in Figure 2, is most commonly selected. However there are also model pilots who swap the left and right sides.

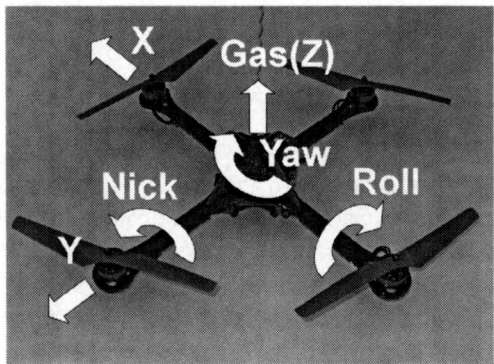

Figure 1: Control capabilities

Figure 2: Stick assignment

'Nick' describes the tilting forward and backward. For that purpose, the stick of the remote control needs to be moved upwards (tilting forward) and downwards (tilting backward).
'Roll' describes the tilting to the left and right. The stick needs to be moved to the left and the right side.
'Yaw' describes the rotation around the vertical axis (z). The left stick needs to be moved to the left (counterclockwise yaw, view from top side) or the right (clockwise yaw, view from top side).
'Gas' describes the movement along the vertical axis (z). If the left stick is moved down, it means descent flight, and if the left stick is moved up into the full throttle position, it means climb flight.

1.2 Physical movement

The immediate question is now how a Quadrocopter can be controlled physically with the above functions. A helicopter will again serve as a comparison.
'Nick' and 'Roll' is realized with a so-called swash plate. This provides at the end an angle-shift of the main rotor force axis to the fuselage. 'Gas' is provided by 'pitching', which is achieved by changing the pitch of the rotor blades. 'Yaw' is realized by a change in speed of the tail rotor. Some models also reach yaw by pitching the tail rotor blades.
Anyone who has ever built and flown helicopters knows that this requires quite a complex mechanism. A hard landing is rarely forgiven: bent rods, ragged ball heads and expensive repairs are the consequence. Many have thus abandoned the model helicopter hobby, the so-called pinnacle of model aircraft.
Quadrocopters, which – as mentioned above – have the same control options as helicopters, in contrast stand out by virtue of their much simpler and thereby massively less sensitive mechanics: There are four motors, which are rigidly connected with two right- and two left-rotating propellers – and that's all.

Everything else is provided by a little electronic control board[1]. Figure 3 illustrates this.

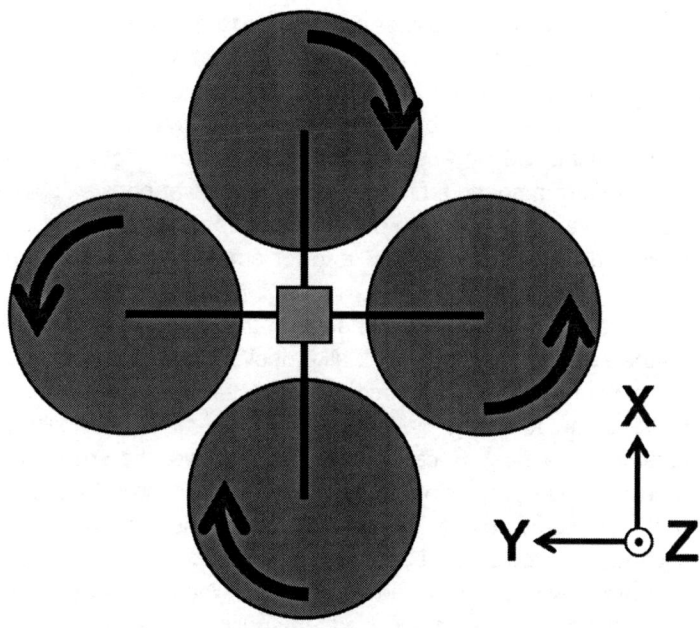

Figure 3: Two left- and right-rotating propellers, view from top side.

'Nick' is physically achieved by a change in speed of the upper and lower propeller (see Figure 3). To move the Quadrocopter in the X direction, the lower propeller is turning faster and the upper one slower. Thus, an inclination in the direction of the x-axis is achieved.

---

[1] A fact which in the technology of the 21st century can be observed very often: A problem that was solved earlier purely mechanically with a sophisticated design is replaced by a combination of sensors, electronics and a much simpler mechanism. Examples as keywords are: mechatronics, fly by wire, direct-drive wheels... the Quadrocopter is therefore in good and modern company.

'Roll' is achieved by a change in speed of the left and right propeller. A movement in the Y direction requires a higher speed of the right and a lower speed of the left propeller.

The main rotor of a helicopter produces a torque about the vertical axis (z) because of its twisting. The tail rotor serves to compensate for that torque. The two right- and left-rotating propellers of the Quadrocopter do this job instead. Thus a tail rotor is not needed. 'Yaw' is achieved by ensuring that both left and right propellers have a different speed than both upper and lower ones. A counterclockwise yaw (viewed from above) requires a higher speed of the upper and lower propeller and a lower speed of the left and right one.

A change in 'Gas' requires a change in the speed of all propellers together. During the climb flight, all propellers have a higher speed.

As mentioned above, Quadrocopters and helicopters are controlled by the same functions and also have the same possibilities of movement – almost. Because of the control over the speed of the propellers it is not possible to fly stably overhead and to 'mow the lawn', as some pilots demonstrate with their pitch-controlled helicopters[2]. However, loops are possible with Quadrocopters. They are flown as with a model airplane, where the gas is taken away close to the apex and strongly accelerated for the subsequent stabilization in the suspense position.

1.3 Flight in '+' or 'x' configuration

Since a Quadrocopter is constructed so perfectly symmetrically, the question of where the front is is justified. Most model pilots think it is as shown in Figure 3 and they mark the front rigger with a piece of tape or similar. This flight configuration is called the '+' configuration.

---

[2] However, there are already prototypes of pitch-controlled Quadrotors. These are special designs, but then the advantages of simple mechanics disappear.

The complete symmetry has its problems though. Thus Quadrocopter pilots seldom fly their flight models further than 50m from themselves, because they will not be able to correctly identify where the front is. The word 'flight model' is also worthy of discussion. 'A model of what?' could be the question of an inveterate model aircraft builder.

Many Quadrocopter pilots have therefore gone over to another attitude of flight, the so-called 'x' configuration. There, the front end is not one of the riggers, but the central point between two of them. There are control circuit boards which support this in the configuration and carry out internally a so-called transformation of coordinates.

Figure 4: In reality a roll bar would probably make sense

However, an 'x' configuration can always be achieved without problems also on the part of the remote control by using the V tail mode, which couples horizontal and vertical tail (here nick and roll).

So the advantage arises that the model can be provided with a body, for example one made of Depron. So, the orientation is always clear, and it is therefore possible to fly larger distances. On the other hand, this brings the Quadrocopter closer to the 'right' model aircraft. Also the creativity in building fuselages is taken into account. Figure 4 shows an example of a model in the 'x' configuration (an example of the future?). More examples and pictures can be found in Chapter 7.

2. Base components

A distinction is made in the following discussion between base and extension components. The issues addressed in this chapter are basic components which are mandatorily necessary for the flight. They are offered by almost all manufacturers in this or similar form. The extension components, which are described in the next chapter, allow additional functions. They can be added as desired. Before the purchase decision you should first consider the application field of your Quadrocopter. Not all systems can be provided with all the enhancements. This should be checked on the Internet. Figure 5 shows the basic components of a modern Quadrocopter. These are described in the following sections.

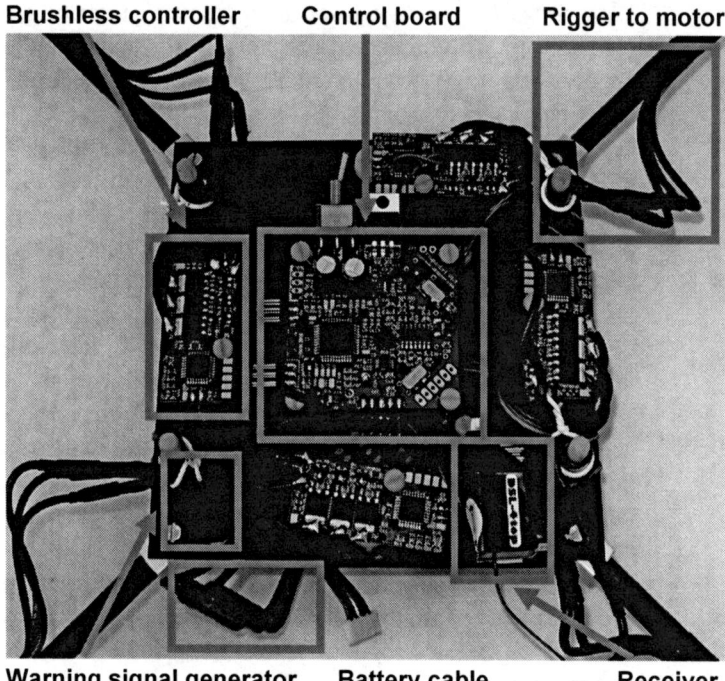

Figure 5: Base components of a Quadrocopter

2.1 Control board

Remote controlled model aircraft or helicopters usually have no control board. The receiver takes over this function. It drives the servos, the gyro for the stabilization of the tail rotor and the brushless controller for the motors. In the case of a Quadrocopter, the functions of the receiver alone are not sufficient to achieve all requirements. This has several reasons.

- The main reason is physics. Without additional sensor support a human being would not be able to stabilize and fly a Quadrocopter. Looking at an axis (such as the pitch axis) from the side, this looks like a swing with two drives at the ends. If both motors do not produce exactly the same thrust (which they never do in practice), the axis will start to turn. To compensate for this, one would have to observe this rotation and manually reduce the gas in one of the motors.
  We can illustrate a similar problem by trying to stabilize a broom which is on the palm of a hand. Balancing it with the hand works pretty well. Now we try to stabilize a 40cm long aluminum rod in the same way. This is much more difficult because the rod is smaller and quicker in its movements.
  It's the same with the Quadrocopter, where the spacing of the drives is usually of a similar size. It is therefore necessary to measure at least the roll and pitch angles. In addition, in the yaw axis, the angular velocity (yaw rate) is measured. The control board then balances the axis through the sensor signals by itself. That means it controls the speed of the motors, but much faster than humans could.
- As mentioned above, several drives together must execute a certain action following a stick movement. It is not the case that a stick is responsible for the movement of only one motor alone. A mixing function is therefore necessary. This is flexible enough only with a microprocessor on the control circuit board and corresponding software.

- The stabilization due to the sensors – the above-mentioned balancing – is realized by PD controllers (more on this in the chapter 'Setting the controller'). The P and D parameters must be adapted according to the choice of other combinations of motors and propellers or larger or smaller riggers. They play the decisive role as far as the properties of the Quadrocopter – agile and fast or slow and good-natured – are concerned. Therefore, it is necessary that they can be reconfigured using the PC interface.
- A configuration with a PC interface allows also a range of other parameters such as the effect of the control stick, an emergency gas function upon a failure of the receiver signal, a battery warning of low voltage, etc.

Therefore, the control board plays a central role in building the Quadrocopters. In summary, it provides the following basic functions:

- Power of the four brushless controllers via a central switch
- Calculation of the angle in the nick and roll axis and the angular velocity in the yaw axis using the gyro and acceleration sensors
- Control of nick and roll angle and the angular velocity in the yaw axis with PD controllers (balancing)
- Calculation of motor voltages due to the control and the throttle position
- Signals for the brushless controllers
- Special functions, e.g. actions after the loss of receiver signal, battery warnings

2.2 Sensors

The sensors are integrated in most Quadrocopters on the control circuit board, including in the example in Figure 5. Since they take a central role, however, a chapter of their own is dedicated to them here.

There is a problem in measuring the angle in the nick and roll axis, and the angle velocity in the yaw axis (using a gyro), so that the control board can balance the system autonomously. Therefore, first the function of the gyro and acceleration sensors will be explained. It is then shown how the angle or angular velocity can be calculated using those sensor signals.

Gyro

Most of the gyros which are used in model construction are based on the effect of piezoelectric elements. With piezoelectric elements, voltage can be applied, whereby it expands mechanically; on the other hand, a mechanical expansion is also converted into a voltage. Both effects are necessary for the operation of a gyro. Figure 6 shows an arrangement in a schematic diagram. There are three piezoelectric elements arranged in a triangle.

To one of them a sinusoidal voltage oscillation is applied. It vibrates mechanically. The so-called gyroscopic effect now ensures that rotation also causes the other two piezoelectric elements to vibrate. The faster the rotation, the greater the vibration, and so the greater the angular velocity. The intensity of the vibration and thus the angular velocity can be measured directly. It is proportional to the voltage of one of the two vibrating elements.

Piezo elements will always have a temperature drift. Therefore, there are various calibration procedures for Quadrocopters with gyros based on piezoelectric elements. More recent (and more expensive) gyros are based on MEMS (Micro Electro Mechanical Systems) technology. They are less susceptible to temperature.

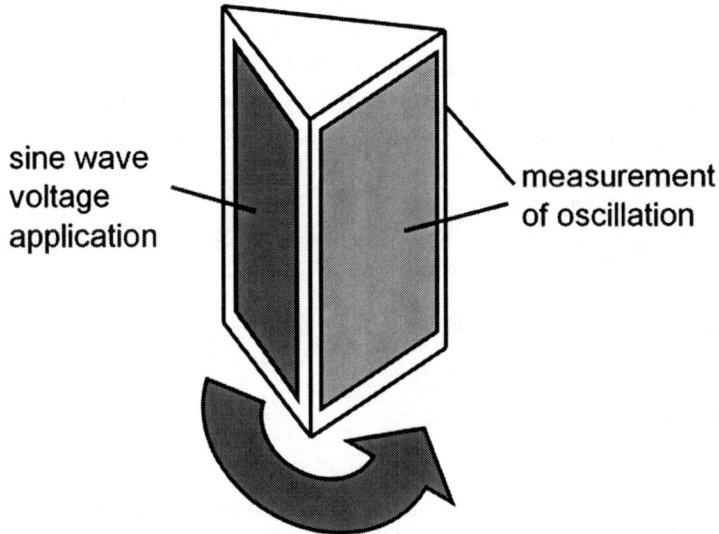

Figure 6: Mode of operation of a Gyro, schematic diagram

Acceleration sensor

Figure 7 shows an acceleration sensor. A mass (known as seismic mass) is suspended by a damped spring in all three spatial directions. If the system is inclined, as illustrated, the acceleration of gravity g causes a shift of this mass.

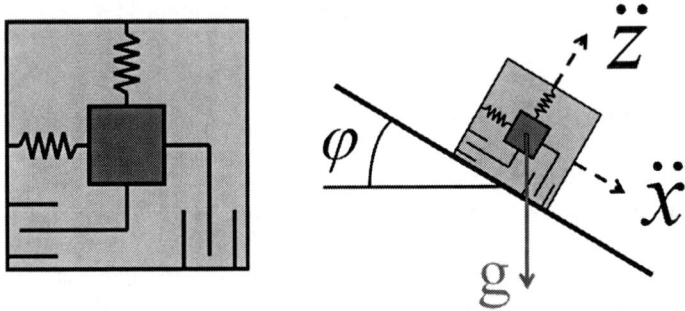

Figure 7: Acceleration sensor

The projected components of g are now measured in the x- and z-direction. The two dots on x and z represent acceleration. The measurement is done capacitively and works so that the moving mass moves a metal plate between two capacitor plates, thereby changing the capacities. Using some trigonometry we see:

$$\ddot{x} = g \cdot \sin(\varphi)$$
$$\ddot{z} = -g \cdot \cos(\varphi)$$

Since 'sin/cos = tan' it follows that

$$\tan(\varphi) = -\frac{\ddot{x}}{\ddot{z}}$$

For a small angle, the tangent is always the same size as the angle itself (to test: set the calculator to RAD) and so the equation simplifies to:

$$\varphi = -\frac{\ddot{x}}{\ddot{z}}$$

Acceleration sensors typically measure three dimensions on a chip, i.e. in the x-, y- and z- axis. The tilt angle in the nick axis, as described above, is measured with the x and z components, and the one in the roll axis with the y and z components.

Combination of sensors

So one could simply say: One needs an acceleration sensor for the nick and roll angle and a gyro for the angular velocity in the yaw axis.
This is correct for the yaw axis. For the nick and roll angle, however, things are a little more complicated. Using both the gyro and acceleration sensor, the calculation of the angles is possible using either the gyro or the acceleration sensor. Both, however,

have disadvantages, so that only a combination of both (called a sensor or data fusion) gives a reasonably reliable result.

A little physics may help: The angular velocity is the time derivative of the angle, in other words explains the angle change per time unit. Summing the signal of the gyro numerically, we therefore form the integral and we get the angle. This is done such that the signals are added at specified times. The problem with numerical integrations is, however, that errors add up and thus after a short time the calculated angle no longer corresponds to the measured angle.

The acceleration sensor does provide the angle, but only in so-called stationary cases. If the Quadrocopter is inclined, an acceleration in the direction of the slope will occur, which is of course also measured. So the angle calculation is invalid. In addition, wind influences distort the result. Only without wind and with a movement with constant velocity (i.e. no acceleration) is the angle measured accurately. Then the above formulas are valid.

A lot of development work in the field of sensor fusion is being undertaken in order to accurately measure the angle with a combination of gyro and acceleration sensors. Model-based approaches or Kalman filters are the technical terms for this. In practice, however, a simple solution is usually implemented. The angle is integrated numerically based on the gyro sensor as described above. The signal of the acceleration sensor, which often – although not always – provides a reasonably good signal for the angle, is used as reference to match the integrated gyro signal. It thus ensures that the angle does not drift away as a result of the numerical integration.

Anyone who runs speed flights with his Quadrocopter has already observed that his aircraft is not exactly straight after quick maneuvers or curves. This is because in this attitude the acceleration sensor can't measure the angle accurately because of the above. The integrated gyro signal is compared incorrectly.

In summary, there must therefore be a gyro sensor in every rotation direction (roll, nick and yaw), and for the adjustment of the angle of roll and nick an acceleration sensor in x, y and z in addition.

2.3 Brushless motors, propellers

Today, brushless speed controllers and brushless motors are state of the art in model construction. This is also very much the case for the Quadrocopter. The first hobby Quadrocopters still worked with DC motors or brushed motors. This technology harbored at least two serious disadvantages.

- Since four drives had to be used, weight reduction was very important. Weight could be saved by simply ignoring the limit of the power. Thus they became very hot and the brush life was extremely short. In a model aircraft with only one drive it is not tragic if this results in only 80% of full power due to the condition of the brushes. The pilot may not even notice the problem. With a Quadrocopter however, a brushed motor at the end of its life cycle means that the other three drives had to reduce speed so that stable levitation was still possible. The four motors were therefore only as good as the weakest.
- DC motors are inherently driven at high speeds. Direct drives were therefore not possible. Thus, a gearbox had to be used, which carried out the matching between motor and propeller. This yielded additional weight.

In summary, by using brushed DC motors much weight had to be carried in relation to the delivered power.
Therefore, it was soon clear that the brushless DC motors used in other areas of model construction are superior to the brushed DC motors. In 2007, the first hobby Quadrocopters with these drives were to be bought.
The operating principle is explained here briefly. In contrast to brushed motors, the commutation of current necessary for turning is not a mechanical contact between rotor and stator (the brushes), but is done electronically. However, special electronics is also needed: the brushless controller. It gives three alternating current signals shifted by 120° on the three cables of the drive.

Since the drive works without mechanical contact, the ball bearings of the shaft are the only wearing part. Many early Quadrocopter pilots often changed the brushed DC motors between flights. Today the brushless DC motors work for many hours, and the lifetime is no longer an issue.

A gearbox can also be avoided in most cases. The brushless motors are in fact offered as inrunners or outrunners. Inrunner means that the outer part is fixed and the inner part is turning, as with brushed DC motors. Outrunners have the rotating part outside, and the fixed part inside.

One of the most important characteristics of a brushless motor is the number of rotations per minute and volt (kV or rpm/V). It shows how many rotations the drive makes per minute depending on the applied voltage at the input of the brushless controller. Taking as an example that there is a 1000 rpm/V motor and 10 volts are applied at the input of the controller, we obtain 10V x 1000 rpm/V = 10,000 rpm. This is at idle running, with no propellers.

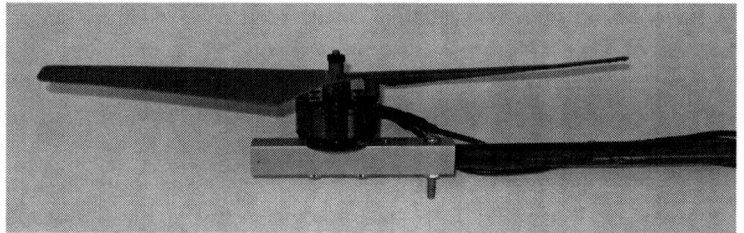

Figure 8: Brushless motor 30g with 8" x 4" propeller

With the characteristics kV or rpm/V, outrunners typically lie deeper than the inrunners – even in an order of magnitude, which makes them attractive for the direct drive of propellers. Many model aircraft fly with such drives.

In the chapter 'Dimensioning of motors and propellers' this is treated a little deeper, but until then the following sentences suffice:

Use motors with a kV of about 1000 rpm/V.
For a motor with 30g weight, an 8" x 4" propeller in direct drive is selected, which gives about 4 x 300g maximum thrust at 10V.
For a motor with 60g weight, a 10" x 4" propeller in direct drive is selected, which gives about 4 x 600g maximum thrust at 10V.

Thrust–weight ratio

Typical constructions of Quadrocopters have a thrust–weight ratio of about 3:1. This means that the maximum achievable total thrust with four motors and propellers is three times higher than its own weight.

If a Quadrocopter with 30g motors and 8" propellers has a total thrust of 1200g, so its own weight should not exceed 400g. At first glance, this ratio appears high. One ought to believe that about half of the thrust should be enough. Experiments show that the Quadrocopter then just barely levitates, if it is windless. This is so because you need a so-called control surplus of about a factor of 2 to have enough power available for the balancing of wind shocks, turbulence, etc. A thrust–weight ratio of 2:1 could keep the Quadrocopter in balance quite well. If the factor is increased to 3:1, either a payload such as the size of its own weight is supported, or speed flight or loops are possible.

Thus, in the interpretation, about 2/3 of the possible thrust is pure reserve. Looking at Figure 8, the motor is in the air stream of the propeller. It is also accessible from all sides. This has among other things the very nice advantage that the drives are usually only lukewarm during hovering and speed flight. The efficiency of the system is increased because the motor thrust mainly produces little heat and, in addition, it contributes to the high life cycle of the brushless motor.

2.4 Brushless controllers

Brushless motors are always operated with brushless controllers. This is necessary because of the special control of the drive with alternating current. They are available for model aircraft from many manufacturers.

They are supplied with a DC voltage input side and, like normal servos, are connected directly to the receiver. The receiver provides them with a pulse width modulated signal[3] of a speed reference value, corresponding to the throttle position of the remote control. They now change the voltage to a three-phase AC voltage using Field Effect Transistors. If a drive needs to turn faster, then the frequency and amplitude of the AC voltage is increased in a way that the speed can follow. The brushless control thus takes the following tasks:

- Evaluation of the speed reference signal from the receiver
- Transforming the supply voltage into a three-phase AC signal
- Starting up the motor speed to the reference speed

Most manufacturers of Quadrocopter systems offer their own brushless controllers. Because the sales volume is not large, they can't compete in price with the standard brushless controllers. But there is a reason why this standard brushless controller often can't be used.

With a flight model, it is not important whether the brushless motor controller accelerates the motor in 10 or 100 milliseconds to the desired speed or whether it does so even after a second. The thrust is then just a little later. However, as described above, the

---

[3] A pulse width modulated signal jumps in the millisecond intervals between two voltage levels, 0V and 5V. With a larger speed reference the value is longer at 5V, and with a smaller speed reference longer at 0V. The receiver communicates in this way with brushless controllers or servos.

brushless controller in Quadrocopters is also responsible for ensuring that the signals from the control board can be quickly implemented in the motor. Remember the example of the balanced aluminum rod. If this is not guaranteed, the stable levitation position is endangered.

The pulse width modulated signals are therefore often too slow a method of communication. This problem can be solved in different ways. Either the brushless controller software can be adjusted, thus increasing the communication rate, or a completely different interface is used.

In various Quadrocopter systems the I2C (read: I-square-C) is set by bus. This is a two-wire line which starts from the control board and connects all four brushless controllers. The transmission of the speed reference is made serially. All brushless controllers are sequentially addressed by a different address (or different sequences of 5V and 0V) and updated with their new speed value.

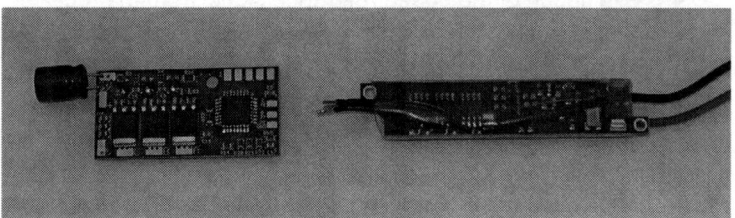

Figure 9: Brushless controller

Capacitor

In all brushless controllers, there is a relatively large (usually electrolytic) capacitor, shown in Figure 9 left. It is connected between the positive and negative voltage supply and smooths out voltage drops when the motor speed requires a higher current for a short period. Its function is also important because those voltage drops would jam the communication with the control board. In a defective brushless control unit it is often the culprit component.

2.5 Lithium-polymer accumulator

Electric drives only became successful with the development of efficient accumulators for model flight equipment. 25 years ago there were only a few gliders with electric auxiliary motors. These were powered by nickel-cadmium (NiCd) accumulators. Later, nickel-metal hydride (NiMH) accumulators came and, in recent years, lithium-polymer (LiPo) accumulators. During this time, the storage capacity of energy per unit of weight has increased by about five times.

As an example, in the middle of the 1980s an electric glider could develop thrust for three minutes with a NiCd accumulator. Today, this same electric glider with an equal weight of the latest generation LiPo accumulator can develop thrust for fifteen minutes. He who dares to look into the future can anticipate that the number of petrol-driven models will decrease further if the batteries become more powerful. Indeed, it is now apparent that the charge capacity (meaning here the energy stored per unit of weight) of the LiPo accumulators can be increased in the coming years by another 50%. Also planned in various research laboratories are new accumulator types with even greater capacity.

This development is also helpful for the Quadrocopter. The author knows of no Quadrocopter project where gasoline engines have ever been used. Since it already takes a lot of patience to get one gasoline engine running, this would be even more difficult with four of them. (Anyone who has watched pilots of two-motor model airplanes, where for example the first engine stops before the second starts, can understand this.)

The LiPo accumulator has some special properties. A cell has a nominal voltage of 3.7V. Usually three or four cells are connected in series, which then results in a nominal voltage of 11.1V or 14.8V. Deep discharging below 3.0V per cell is not liked by the accumulator; hence the use of battery warning systems. These can usually be configured on the controller board via a PC interface, whereby below a certain voltage a warning via a flashing light or buzzer sound occurs.

The charging is also carried out in a way that NiCd or NiMH batteries are not used to. An example of a three-cell LiPo accumulator, in which the battery alarm is set to 9.0V, will show this. Assuming that one cell has 3 volts, and the other two have 3.3V each, it is clear due to the resulting sum of 9.6V that the accumulator warning is far from being activated. The battery seems almost full. If you continue flying, the warning is activated at 3 x 3V = 9V. Then the first cell has about 2.8V and the other ones 3.1V each. Then the 'problem cell' is already heavily discharged and its lifetime is greatly reduced.

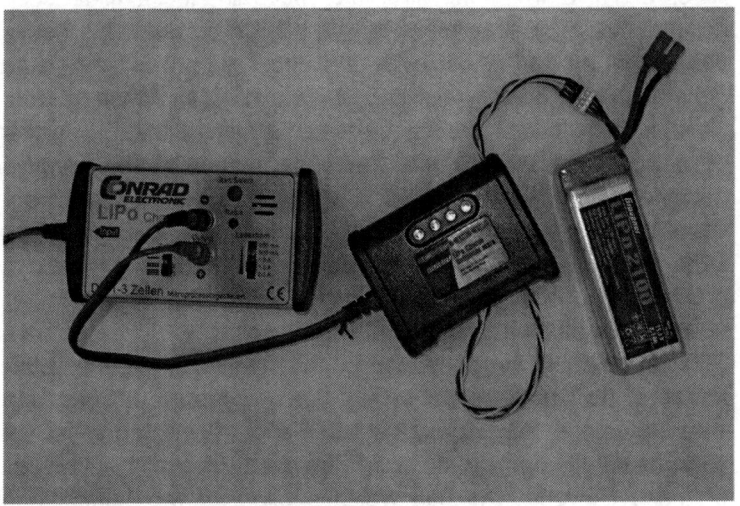

Figure 10: Recharger with balancer and LiPo accumulator

Multi-cell accumulators are today loaded with a so-called 'balancer'. This is a device that is connected between the charger and accumulator. It ensures that all the cells have the same voltage after loading. The danger of too low a voltage of the cells is largely prevented. For this, another plug is required, which leads the positive and negative poles of all cells wired in series. Overcharging – a charge to too high a voltage – is also not desirable, and this is also controlled by the charger and balancer.

Deep discharging and overcharging breaks down the anode, creating hydrogen. This inflates the cell. A swelling of the cell is a sure indication that the LiPo accumulator should be replaced. Also in the event of mechanical damage, dents, and the like, a new LiPo accumulator pack should be used.

Dimensioning

It depends on the total weight, the payload, the motors and the desired flight time how large the accumulator should be. Therefore some considerations should be made.
As a simplification, with the use of 8" or 10" propellers about 100W power per kg of mass levitation is expected[4]. If a flight time of 20 minutes is assumed (for speed flight, this is admittedly somewhat reduced), then for the two examples with a three-cell LiPo accumulator the following results are reached:

Motors with a weight of 30g and 8" x 4" propellers and a total of 4 x 300g = 1200g maximum thrust and a Quadrocopter with a total weight of 400g (thrust–weight ratio of 3:1):
100W/1000g x 400g = 40W power; 40W/11.1V = 3.6A current for levitation,
Charge content of the battery: 3.6A x 0.33h = 1200mAh

Motors with a weight of 60g and 10" x 4" propellers and total of 4 x 600g =2400g maximum thrust and a Quadrocopter with a total weight of 800g (thrust–weight ratio of 3:1):
100W/1000g x 800g = 80W power; 80W/11.1V = 7.2A current for levitation,
Charge content of the battery: 7.2A x 0.33h = 2400mAh

---

[4] This is valid for the pure floating. If larger propellers are used, less power per weight is needed. Therefore helicopters also have large rotors. In Chapter 6, 'Dimensioning of motors and propellers', this subject is treated in detail.

Of course, the total weight of the Quadrocopter also includes the weight of the accumulator; a three-cell 1200mAh LiPo accumulator today weighs around 100g and a three-cell 2400mAh LiPo accumulator 200g.

2.6 Remote control and receiver

The remote control is for many model constructors one of the most important devices. Here they set the properties of their model. The aileron servos or V-tail are configured, the throttle is implemented with an exponential function, etc. 'Why is this so?', one might wonder. Because in a typical aircraft or helicopter model in fact only the receiver is fitted and this has only limited or no configuration options at all. The modern remote controls therefore have memory, so that they can be programmed for several different models. If necessary, it can easily be changed from one model to the other by pressing a button.

With Quadrocopters, it is a different, 'inverted' world. The control electronics also includes a microprocessor. Thus, the parameters can be downloaded from a PC interface. This means that there are no special functions necessary for the remote control and that the whole configuration is stored on the microprocessor of the model.

Unlike the model airplane and helicopter pilots, who configure the remote control for the model, the Quadrocopter pilot configures the model for the remote control. Parameters such as exponential functions, leverage, etc., are adjusted via graphical user interfaces by the PC and then downloaded to the control board via the interface.

Requirements

So, as shown above, model memory is not necessary. The more easily you can use the remote control, the better it is. In order to fly x-configuration (see Section 1.3), sometimes the possibility of a V-tail unit is needed, but that's about it. The question of the number of channels is more difficult to answer. The minimum is the four channels of nick, roll, yaw and throttle. Depending on how much you use the extension components considered below, even more channels will be needed. Most of the control electronics therefore support additional functions for another two or four channels. So, the remote control system should be equipped with 6 to 8 channels in total.

2.4GHz

Today, the author would also advise 2.4GHz technology to a new entrant, even though he has been an avid 35MHz (Germany) pilot for many years. It does have some compelling advantages so that, in a few years, it will probably have replaced the 35MHz (Germany) or 72MHz (USA) controls. Through the so-called 'frequency hopping' it is no longer necessary to hold consultations on the airfield on which channel is allowed to fly.
The system simply 'hops' (jumps) through all possible frequencies in the permitted frequency band at 2.4GHz and on one of these the signal reaches with (almost) 100 percent probability the receiver. Furthermore, the technology is bidirectional. In various systems the signal strength is displayed directly on the transmitter. This is a very simplified explanation. For more information on 2.4GHz technology, refer to the literature.

35MHz (Germany) / 72MHz (USA)

Each country has its own laws which frequencies for model aircraft are allowed. As an example, there is 72MHz in the USA and 35 MHz in Germany (2011).
For Quadrocopters in most countries, the same channels are allowed as in model airplanes and helicopters. Since in some countries there are differences, you should check the legislation of your country. In many systems, the receiver is not integrated into the control electronics. For some, the communication interface

between the control board and receiver is the so-called PPM-sum signal. This is a signal that transfers eight channels serially.

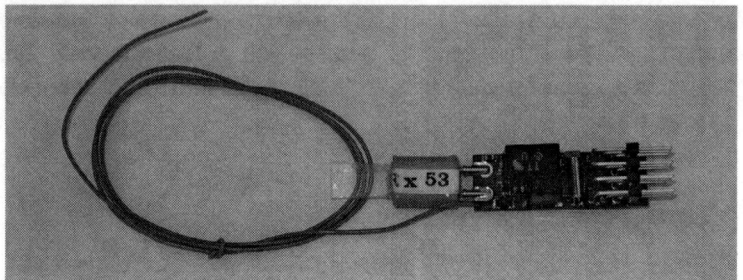

Figure 11: Standard receiver, version with ppm signal

2.7 PC interface

Since almost everything can be configured with the PC, a communication interface is necessary. Figure 12 shows two variants from different manufacturers. On the PC side, it is a USB interface. The signals are converted into a signal compatible with the control electronics. The PC then runs a configuration program which is downloadable from the Internet.

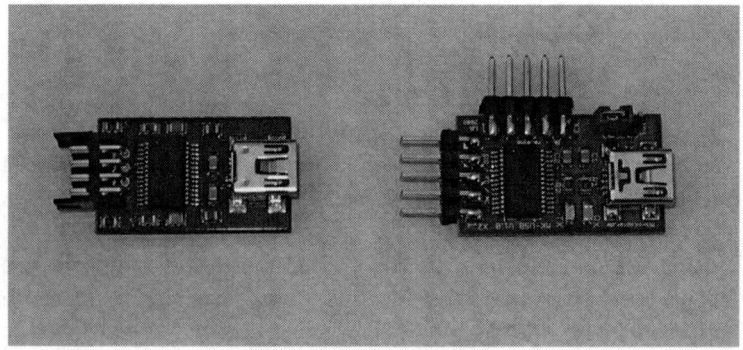

Figure 12: PC interface

It offers the possibility to set the parameters and also has specific diagnostic capabilities. As an example, you can watch the measured sensor data, or the signal of the remote control. It is also possible to allow the start of drives, etc. When troubleshooting, this is an excellent tool because it figures out pretty quickly what is not working.

2.8 Frame construction

The frame carries all components, motors, propellers, control board, brushless controllers, battery and receiver. It is often underestimated. It has the same function as the fuselage in model airplanes.
With a Quadrocopter which does not show the required behavior in flight, the frame is very often the cause of the problem. This may be because the drives are not sufficiently rigidly mounted, the control electronics with the sensors vibrates, the antenna cable is wrapped around the rotary motors, etc. It will also be pointed out here what one has to consider with the purchase or construction of the frame.

Crash-safe or with predetermined breaking point

Depending on the field, the author would distinguish between the two frames in the title. For a speed flight oriented pilot, a crash-safe framework would be recommended. Such frames can be bought in several online shops. They are usually drawn with a CAD program; sometimes the forces to be borne were simulated and tested. Often, a material is used which until recently was reserved for the professionals: carbon. It is characterized by being very strong but also very light. So it is ideal for these purposes.
Figure 13 shows such a framework. The electronics and battery are protected by a metal case, whereas the rigger consists of a sandwich with balsa wood glued between two carbon plates. The

brushless controllers are mounted under it. The price of crash-safe frames tends to be at the upper limit.

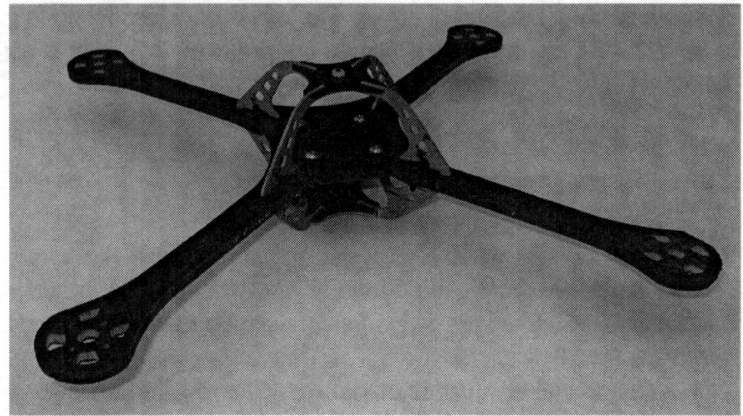

Figure 13: Crash-safe frame

Frames can also be less crash-safe, if they are at least rigid. In this category are the so-called do-it-yourself shop frames. They get their name because you can find almost all the components for the construction in the DIY home-improvement store. The hardware of such frames, however, should be strictly engineered so that if something breaks, no expensive electronics or drives must be replaced. This means that they need a predefined breaking point.

Frame size

The frame sets the size of the Quadrocopter. An important indicator is the distance between two opposing motor shafts. Often we hear the terms Quadrocopter of the 30s, 40s or 50s class. The numbers denote the spacing in centimeters. There are only geometric constraints that determine them. Care must be taken to ensure that the propellers can rotate freely. But it is not necessary that for a smoother balance, for example for a camera flight, that the distances need to be as large as possible. Only the quality of the sensor measurement, the quality of the motors and propellers, and the controller parameters determine this. Simple rules can be created for the frame size as well:

8" propellers will need 30s to 40s class frames,
10" propellers will need 40s to 50s class frames,
12" propellers will need 50s to 60s class frames.

It always also depends on where the brushless controllers are arranged. With most crash-proof purchase frames, the riggers are so big that there is also room for the controllers. So in this case the center will be made smaller, since it only needs to carry the control board, the receiver and the battery. The distance between two opposing motor shafts is then smaller. With a DIY-store frame, the riggers are usually made of aluminum or carbon tubes, which may not carry any additional electronics as they would otherwise not be sufficiently protected. Then the center has to be designed larger, as also the brushless controllers have to find a place (see Figure 5).

Construction of a DIY-store frame

The hobby constructor of a Quadrocopter needs to buy electronics, motors and propellers anyway. Thus, only the frame has got a real potential to save money.
For this reason, the construction of a relatively cheap DIY-store frame is shown, which is very rigid for all sizes and has a breaking point in the case of a crash. So it protects the electronic components and, moreover, it looks pretty good.
The center plate of 130x130mm must carry the expensive electronic components and the accumulator. It is therefore important to ensure that it does not suffer any damage in a crash. So, first, an aluminum cross is built according to Figure 14. The square tube has the outer dimensions 10x10x185mm and the interior dimensions 8x8 and can carry a carbon tube of 8mm diameter.

In the middle, an 8mm hole was drilled to hold an 8mm aluminum tube. This is attached to two 10x10x85mm square tubes and screwed together. 10mm from each edge, a 3mm hole is drilled. Later on, the riggers will be screwed into them.

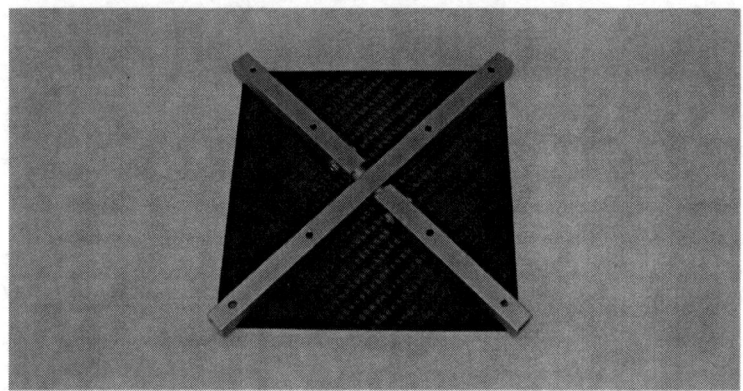

Figure 14: Center plate, aluminum cross

Now two 130x130x1mm carbon plates are cut, which are bolted to the aluminum cross with M3 nylon screws. The distance of the holes from the center is 50mm each. Since the carbon plates are relatively expensive, 0.8mm aluminum plates can be used instead. However, there are then some restrictions, as a higher weight and a lower strength is the result.

The accumulator is on the bottom side and is held with M6x40mm nylon screws at a distance from the base. For the protection of the electronics, carbon pieces are used together with the same nylon screws. As spacers, carbon tubes 8mm in diameter are fitted. Figure 15 shows this.

The riggers are made of 8mm carbon tubes; alternatively and with the same restrictions as above, aluminum can also be used. Many DIY-store frames whose riggers are designed with tubes are not rigid enough and the drives can quite easily be twisted. The manufacturers of electronics warn, as in this case, that no reliable control of the system (no reliable balancing) is guaranteed.

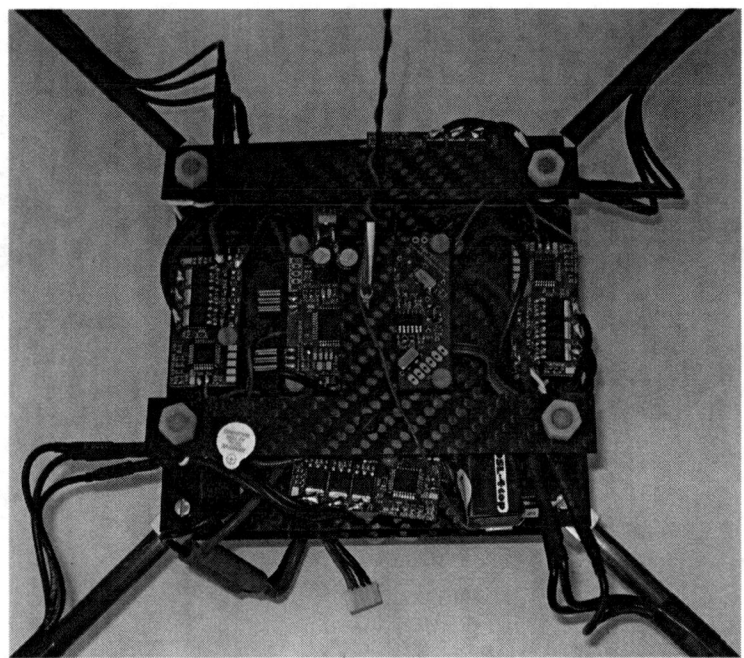
Figure 15: Center plate with electronics and riggers

The version presented here solves this problem. At the end of the carbon tubes are again 10x10x70mm square aluminum tubes for the drive mount, bolted with M3 screws. In addition, two riggers are plugged together using other carbon tubes (4 to 6mm) and smaller aluminum square tubes. This is for those who are flying in 'x' configuration (see also Figure 4 again). If flying in the '+' configuration, for reasons of symmetry all four riggers should be connected with four 4mm carbon tubes. Of course this creates an even greater rigidity with a simple and relatively lightweight solution.

This additional connection provides on one hand for sufficient rigidity and ensures that the drives can't move against each other, and on the other hand also ensures that all motor shafts are vertical. Figure 16 shows a completely screwed rigger; it is a detail of Figure 4.

Figure 16: Completely screwed rigger using a 60g motor

In a crash, this framework is certainly not as crash-proof as a purchased frame, which is designed for speed flight. However, you will always have the 8mm carbon riggers to break and furthermore exactly where they exit from the square tube of the middle piece, so there is a breaking point. This is very desirable for two reasons: The expensive electronics lie safely on the unbreakable center plate and the deformation energy of the crash does not act on the motors and propellers. Most importantly, only a relatively cheap part, the 8mm carbon tube, which is attached with two screws, needs to be replaced. In five minutes the damage is repaired on the airfield and the flight can continue. From 1m carbon tube more than one replacement tube can be manufactured. Depending on the size of the Quadrocopter, their length is from 15 to 25cm.

2.9 Cabling

Many manufacturers describe the wiring of their products on their website. At this point, some additions should be made, as it represents an important issue for the reliable operation of the Quadrocopter.

The wire cross-section should have at least the following sizes, assuming 30g motors / 8" propellers (60g motors / 10" propellers):

Accumulator – control board: $0.75mm^2$ ($1.0mm^2$)
Control board – brushless controller: $0.75mm^2$ ($0.75mm^2$)
Brushless controller – motor: $0.5mm^2$ ($0.75mm^2$)

The wiring must be performed with a star shape, which means that each brushless controller gets separately two (+/-) power supply cables from the control board or directly from the accumulator.

The data lines of the I2C or pulse width modulation, which transmit the rotation speed target values from the control board to the brushless control, should be placed as far away as possible from the power supply cables.

Connectors between brushless controller and motor

To ensure the interchangeability of the drives or to allow a simple disassembly for transport, it is sensible to use connectors between the brushless controller and the motor. In the above-discussed frame construction, connectors with 2mm gold contacts were used. This allows an easy dismounting of the entire system. After removing the four screws on the middle piece and the four screws at the end of the riggers, the entire Quadrocopter can also be transported in a small bag. Figure 17 shows a huge Quadrocopter with 80 cm spacing of the motors and 20" propellers.

Where to put the antenna cable

The antenna cable is almost always in the wrong place if so many propellers are turning in tight spaces. There is really only one method for the placement: You should wrap it around a rigid holder and poke it up in the middle of the Quadrocopter. If possible, the holder should not be electrically conductive, and not made of carbon, as this will reduce the range of the remote control. The time saving for a poorly installed antenna cable never pays off. Sooner or later it will be hit by a propeller. Figures 4 and 17 show examples of Quadrocopters with 35MHz antennas. If one uses a 2.4 GHz system, the receiver antenna is very short. The good range of the remote control is ensured when it is brought out as far away as possible from the conductive carbon.

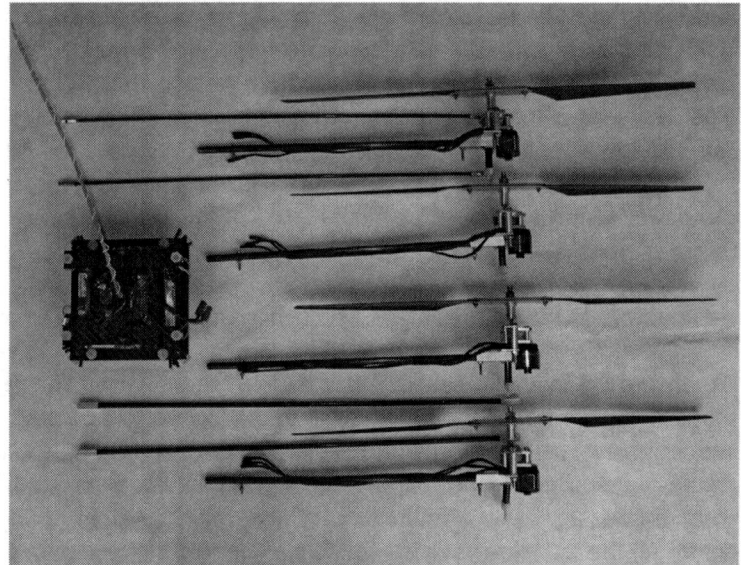

Figure 17: Dismounted Quadrocopter with 8 removed screws

2.10 Safety

The technology is interesting, but it is also important to consider safety. As with all helicopter-like aircraft, Quadrocopters also have propellers which are mounted in the horizontal. On top of that, the used propellers are often very hard. Accidents caused by uncontrolled Quadrocopters falling on people could have dire consequences.
Safety is one of the basic components – it is not optional! Opportunities for this are rings made of styrofoam or, as pictured,

carbon fiber. Here, 0.5mm-thick plates were used. In the concrete example they can be realized with a weight gain of 120g.

Figure 18: Safety ring

3. Extension components

With the basic components the Quadrocopter pilot already holds in his hand a system with which he can perform, as described above, the same control features as with a helicopter. He can on the one hand operate speed flight by attacking the model at an angle, thus generating a driving direction. On the other hand, he can also levitate the model right on the spot. Well-trimmed, with the angle control of the roll and nick axes and the angular velocity control in the yaw axis, in still conditions it is perfectly possible to let go of the remote control and watch the model slowly drift away. And yet there are other ways to control it even better, with additional sensors. To illustrate this, Figure 1 (now Figure 19) is shown again.

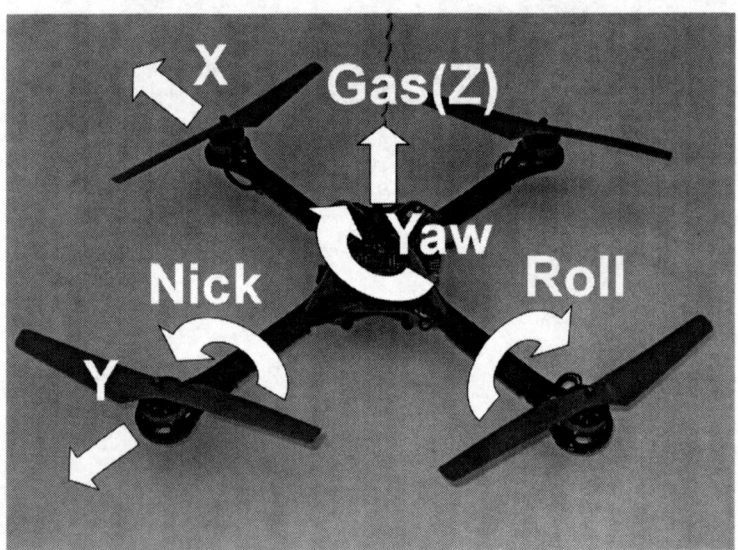

Figure 19: Six degrees of freedom of a body in space

Generally a body in space has six degrees of freedom. In this way, the possibilities of movement are defined. As shown in the figure, there are the three degrees of freedom of angles (nick, roll and yaw) and the three degrees of freedom of axes (x, y and z).

Using only the basic components, relatively few degrees of freedom are actually controlled by the model itself. These are only the two angular degrees of freedom nick and roll[5]. The fun of flying is therefore due to the fact that the Quadrocopter pilot can steer his system to wherever he desires. This means technically that he can determine the yaw angles x, y and z by himself.

However, with today's technology it is possible to measure and control all six degrees of freedom through the Quadrocopter itself. It is therefore capable of staying stable in one place in the air, and the dream of an 'air nail' or 'air anchor' becomes true.

The following section describes briefly which sensors are needed to measure and control the desired degrees of freedom. The individual sensors are then discussed in their own subsections.

- Air pressure sensor: Required for the degree of freedom in the z-direction.
- Compass: Required for absolute angle measurement of the yaw angle (the gyro for the yaw angle measures only the angular velocity, as described in Section 2.2).
- Global Positioning System (GPS): Required for the two degrees of freedom in the x- and y-directions.

These four degrees of freedom can be added to the two degrees of freedom already measured and controlled with the basic components (the nick and roll angle), resulting in these six.

Anyone who is undecided in what order he should expand his Quadrocopter (if desired), is suggested the following:

Firstly air pressure sensor, secondly compass, thirdly GPS

The price of the sensors also increases in this order.

---

[5] Actually, there are 'two and a half', since the absolute angle is not controlled in the yaw axis, but only the angular velocity. The compass of Section 3.2 will solve this problem.

Before the explanation of the extension components, it should be pointed out that not every Quadrocopter pilot wants to equip his system with all sensors. Depending on the application, it may well make sense to use only the basic components. More technology support with additional controlled degrees of freedom also means that he has fewer degrees of freedom that can be controlled by hand.

A Quadrocopter controlled in all six degrees of freedom and stationary in the air is, from the standpoint of technologists and camera pilots, seen as the ultimate goal, but from the standpoint of speed pilots however nothing more than fun braking.

3.1 Air pressure sensor

Measures the altitude and ensures its control (z-degree of freedom).

Figure 20: Air pressure sensor

Air pressure decreases with increasing altitude. A reference pressure pushes on one side of a membrane and the air pressure pushes on the other side. If this changes, it distorts the membrane and a strain gauge measures the electric voltage. Figure 20 shows an air pressure sensor. The air flows through the hole in the middle in the direction of the membrane.

Not only a change in altitude results in air pressure changes in the sensor. Other factors, such as the air stream of the propeller or wind bursts, can also cause this. This must be taken into account during the assembly; placing the sensor directly on the control board with the hole facing upwards or downwards is appropriate. In order to shield the sensor from further disturbances, we can glue some foam or attach tape with a needle hole onto it. But it is important when doing this to ensure that the membrane is not damaged.

Control of altitude

There are several ways that the control of the height can be realized. Two possibilities will be presented below.

In the first variant, if the throttle stick is in the middle position, a zone is defined in which the Quadrotor holds the reached altitude. It then adjusts the thrust by itself. If the throttle stick is above this zone, climb flight will be achieved, and if it is below, descent flight.

In the second variant the throttle stick must be moved so that the system passes into a climb flight. If a lever on the remote control is now moved, it holds the current altitude and controls the thrust by itself. If the gas is withdrawn, it turns into a descent flight. If afterwards the position of the throttle stick is increased again, the system returns to the climb flight. As soon as the height at which the lever was moved is reached again, it again holds that height. The control of the height therefore works metaphorically as an 'umbrella' above which the Quadrocopter can't fly out.

In both variants, the height control can be switched on and off with an additional remote control channel. This has the advantage that the model pilot can decide for himself when the Quadrocopter regulates the height by itself and when he has full control of the throttle stick and the thrust of the propellers.

There are also other options for the realization of the height control, but it would be beyond the scope of this book to describe them.

3.2 Compass

Measures the absolute angle in the yaw axis and ensures its control (yaw degree of freedom).

If only the base components are used, the gyro in the yaw axis only measures the angular velocity (and not the angle itself). When the yaw stick of the remote control is now held in the middle position, that means angular velocity 'zero' for the control board. This also means that the yaw angle will not be changed. The 'nose' (i.e. the front, as far as one can speak of 'nose' and 'front' with regards to Quadrocopters) always remains in the same place. This is true in theory at least.
In practice, the gyro measures an angular velocity of zero even if this is in fact not exactly zero. Because of temperature effects this can't be prevented. A gyro calibration before the flight also does not help.
Thus, the 'nose' will turn away more or less slowly, depending on the quality of the calibration or trimming. The model pilot compensates for that in the same way as with helicopters, namely by a counter-movement of the yaw stick or the trimming.
If it is desired that the Quadrocopter always holds the nose at the same place at the yaw stick position zero, an additional sensor is needed, namely a compass. This will no longer measure a relative motion to the ground, but the absolute orientation.
Many electronic compasses operate on the basis of Hall sensors. The Hall effect is the occurrence of an electrical voltage in a current-carrying conductor which is placed in a magnetic field. The earth produces a magnetic field between the North and South Poles, which is also responsible for ensuring that the compass needle always aligns to the north. The word 'always' should to be treated with caution. If you are in a closed room with metal in the walls or in an area with power cables or iron-rich rocks then the compass needle can be deflected.
The electronic compass is also affected by such disturbances. It is therefore important that they are well shielded or suppressed by other appropriate measures. So the compass should be mounted as far away as possible from the control electronics, the brushless

control and the piezo buzzer. Because of metal in the walls, problems with incorrect measurements are often also reported with indoor flights. Since the built-in compass moves on the nick and roll movement together with the Quadrocopter, it must be designed in a tilt-compensated manner, otherwise it will also result in errors.

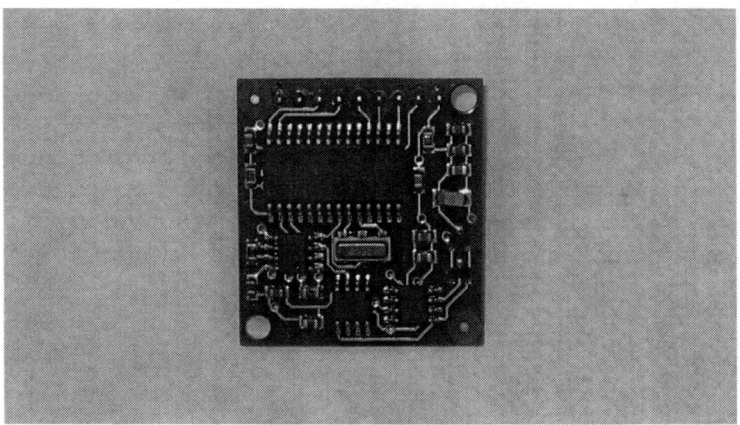

Figure 21: Compass

Control of the yaw angle

As with the realization of the height control, there are also different options for the regulation of the absolute yaw angle. A simple solution is similar to one described there. If the yaw stick is in the center position, or it deviates only slightly from it, then the angle control is active together with the compass and the current absolute angle is kept. If the stick position is sufficiently far enough left or right, it is regulated using the gyro sensor as in the configuration with the basic components. This means that the farther left or right of the center position of the stick is, the faster it turns the nose.

The compass (extension component) and the acceleration sensor (basic component), which measure the yaw angle and the nick and roll angles, respectively, have an interesting commonality. With both types of sensors, the angles are measured in the world coordinate system. In other words, using those sensors, the angular position of the Quadrocopter is always measured relative to the Earth. Thus, the sensors must also be based on earth-related measurement principles. In the case of the compass it was mentioned above that it measures the magnetic field between the North and South Poles of the Earth. With the acceleration sensor it can be seen in Figure 7 that the gravitational acceleration acts as a reference there. It always points to the center of the Earth, so it is also earth-related.

Without the GPS described in the next chapter, all three angle degrees of freedom can be controlled with the sensor types discussed above: nick, roll and yaw. Furthermore, the z-axis (gas) can also be controlled using the air pressure sensor. In total that corresponds to a control of four of the six degrees of freedom.

3.3 GPS

Gives the world's coordinates in longitude and latitude and allows the Quadrocopter to be regulated in the x- and y- axes (x and y degrees of freedom).

GPS, Global Positioning System, has affected the everyday life of people for many years. Electronic navigation systems are now standard in many vehicles. There are simple technical solutions, providing a GPS signal on a single chip.

The GPS consists of several satellites orbiting the earth. Their exact position is known at all times. Each satellite is equipped with an atomic clock and sends out a signal at certain times. That signal also contains its identification. The ground station, a navigation device or, as shown in Figure 22, a chip on the Quadrocopter, measures the running time of the transmitted signals. This is the time which passes between the emission of the signals through the satellites and their reception. In this way it can determine its own position.

The running time multiplied by the speed of light is the distance from the satellite. To locate a particular place on earth, the signals from at least three satellites should be visible at the same time. To measure the exact duration, the ground station would also have to be equipped with an atomic clock, so that all the devices use the same time base. Because that would be much too expensive, in practice such a clock is not installed at the ground station. In order to nonetheless determine the position, the position signal of a fourth satellite is needed. Thanks to special algorithms, the ground station can then calculate the longitude and latitude by itself.

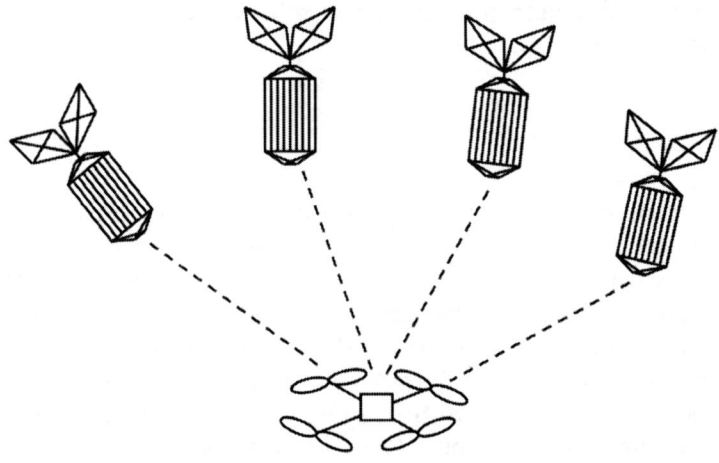

Figure 22: Global Positioning System, GPS

Legislation

It should be noted that a system operated in this way is no longer a pure model aircraft as in the usual sense. It is part of the core tasks that you can steer it with the remote control in different directions. A system regulated in all six degrees of freedom takes over all control tasks; the model pilot no longer needs to move the stick of the remote control transmitter. The Quadrocopter is therefore a drone. Also due to the progressive spread of the miniaturized chip-size GPS receiver, the number of such unmanned and autonomous systems in the air has increased. So many state governments issue guidelines on their use. This is also necessary because the systems represent a security risk.

As a minimum requirement the user must be able to ensure that, any time, he can take over full control via the remote control again. However other laws can also be enacted, for example that eye contact must always be maintained between user and drone or that the pilot must be registered with the government. In some states, this form of control could also be completely forbidden.

Each Quadrocopter pilot must become familiar with the laws in his state and comply with them.

Control of x and y

With the simplest version, the GPS function is turned on and off with the flip of a switch on the remote control transmitter.

If it is turned on, and the nick and roll stick is in the center, the Quadrocopter controls the nick or roll angle by itself, to maintain its position in the x and y directions. If the stick is deflected far away enough from the center, the pilot regains full control of the angles.

The signals from the GPS can only be measured in the open. Thus, the control by x and y doesn't work indoors.

Broader technical possibilities

Thus, if all sensors – gyros, acceleration sensors, air pressure sensor, compass and GPS – are turned on, the Quadrocopter remains more or less as if rooted in the air. All six degrees of freedom of a body in space will be controlled autonomously.

The technique, however, has no limits. For example, the remote control transmitter can be connected to a portable PC and, with the appropriate software, so-called way points can be programmed and those points will be flown to autonomously. Even for such tasks, there are already various purchasable solutions for hobby and professional use.

3.4 Onboard cameras, video-recording

One of the motivations for the development of Quadrocopters in the early 2000s was the prospect of producing photos or films from new, almost arbitrary perspectives. The first professional users were police and fire departments, who were able to photograph scenes of accidents or fires from any angle. Today, camera inspections are undertaken for many different purposes, such as inspecting tall bridges to determine corrosion damage or cracks which are not visible from below.

Figure 23: Onboard camera mounted on top of a Quadrocopter

Onboard cameras

There are countless cameras on the market which are suitable for installation on a Quadrocopter. One must be aware of two things above all: the weight and the ability to electronically trigger images. As a rule of thumb for weight, it is assumed that the camera may be a little more than half as heavy as the weight of the

Quadrocopter without camera. Here, a thrust–weight ratio of 3:1 (without camera) is subject of the condition.

For a Quadrocopter weighing 400g without a camera, the camera must have a maximum weight of 250g.
For a Quadrocopter weighing 800g without a camera, the camera must have a maximum weight of 500g.

Thus, the thrust–weight ratio of 3:1 is reduced to something less than 2:1. There is still enough power there to take photos while hovering.
Many Quadrocopter control board manufacturers provide technical solutions for remote triggering. Of course, this also requires a channel on the remote control. Today's cameras have very large capacity memory cards. One can often therefore do without an automatic picture triggering and program the camera to take pictures automatically at predefined intervals. This works on the condition that its settings allow that. Even video recordings can be realized in this way.

Camera holder

The construction can be carried out in various complexities, according to the need. In the simplest variant as shown in Figure 23, the camera is screwed on so that the lens is aligned between two riggers. This has the advantage that, besides a tripod screw, no special holding device is needed. A disadvantage is the poor protection from contact with the ground in the event of a crash. On the other hand, the view angles are set rigidly before the flight. This greatly limits the perspective possibilities.
A more complicated variant is the pivoting suspension. The camera can be moved here with a servo, as for example shown in Figure 47. Even more sophisticated is a gimbal, which allows for a tilt in two directions. Most of the control electronics have additional servo outputs, which can be configured through the PC interface. Of course, that also requires additional remote control channels.

Swivel-mounted camera holders are often designed so that they are below the Quadrocopter and offer the camera some protection. Since one usually photographs or films downwards, in this way no propellers or motors disturb the picture. But the extra weight is to be kept in mind. Anyone wanting to build a Quadrocopter for such purposes needs to build bigger in order to have a larger payload:

The size of Quadrocopters for flights with swivel camera mounts should be at least:
60g motors, 10" propellers, 50's class frame, weight 800g without camera + mount, 2400g maximum thrust, 2400mAh battery.

Bi-directional connection, video glasses

On the basis of 2.4GHz, there are video cameras that transmit the image wirelessly to a PC. They are also able to receive commands from it. Lighter-weight versions of these devices are ideal for use on Quadrocopters. It is possible to start a video recording from the ground by sending commands from the PC, and to present the film and directly received image data on the screen and to save it to the hard disk.
A strange combination is 'video glasses and Quadrocopter'. Video eyewear works as above, only the video signal is not transferred to the PC but to a display incorporated in a pair of glasses. A flight by video glasses has already been done by some Quadrocopter pilots. One can thus visually feel as if you were standing directly on the command bridge of a Quadrocopter.
The time delay of the received signal is strong with many video glasses. This is for the normal intended use in multimedia applications not usually a big disadvantage. But if you want to control a Quadrocopter, it is essential to ensure that the delay is not too large – as a guideline not more than 0.2 seconds – as the control would otherwise not be possible.

4. Flight mechanics

To get straight to the point: You can be a passionate Quadrocopter builder and pilot even if you have no interest in physics and in what is treated in the next three chapters 'Flight mechanics', 'Setting the controller' and 'Dimensioning of motors and propellers'. The understanding of the matter is still ensured if you read only the findings in italics at the end of these chapters.
For anyone who wants to understand why a Quadrocopter flies at all, or how it realizes the self-balancing, and how to choose controller parameters so that it reaches a predetermined angle of attack in nick or roll, or what motors and propellers one uses, however, the reading of these chapters is recommended.

4.1 Hover flight

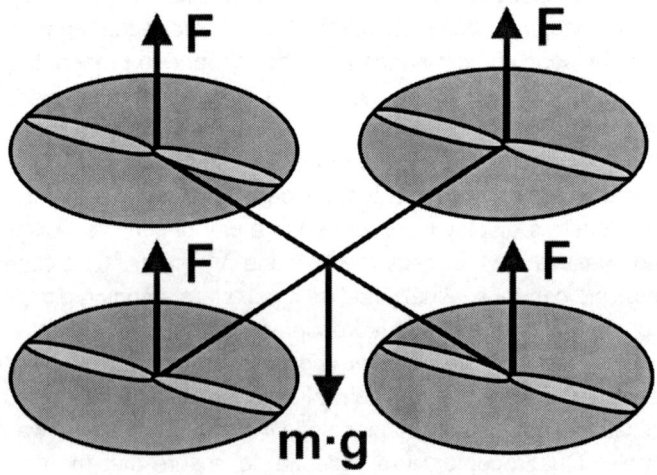

Figure 24: Forces in hover flight

Figure 24 shows a force diagram for normal hovering. From physics 'actio and reactio' is known, and the same can be applied here. In hovering flight namely the sum of the four upward forces is equal to the downward force. The upward-pointing forces F are created by the four propellers and called 'thrust forces' and the downward-pointing force is the weight force (mass m multiplied by the acceleration of gravity g). So:

$4 \cdot F = m \cdot g$

In this equation, something about the units should be mentioned. Usually the model pilot declares the thrust force in grams or kilograms, and this book follows this approach. But strictly speaking, the force should be stated in the units of Newton. The value is simply the thrust in kilograms multiplied by the acceleration of gravity g (about $10m/s^2$). A Quadrocopter which produces 1.2kg of thrust generates, expressed physically correct, 12N of thrust.
In this way, the equation above is also correct. On the left side is for this example 4 x 3N, on the right 1.2kg x $10m/s^2$. To compensate for the 12N weight force, each of the four drives must reach 3N thrust force (or in the language of the model constructor 300g or 0.3kg thrust).
With slight modifications of this formula climb flight or descent flight can also be treated. For the climb each drive produces more than 3N thrust, and for descent less.
Climb flight: $4 \cdot F > m \cdot g$, descent flight: $4 \cdot F < m \cdot g$

4.2 The attack angle φ

So far so good. Next, the pilot would like to attack his Quadrocopter at an angle in the nick or roll direction, so it can drift off to one side and movements in the x- or y- direction are possible (recall the degrees of freedom in Figure 19).
How a rotation by the angle φ is actually achieved is looked at later. For the time being, only the forces for the Quadrocopter with attack angle φ are considered. Figure 25 shows this from the side. It is the same whether nick or roll is shown, because of the symmetry. Of course, even now the four thrust forces are still

active. The middle two shown are from the two propeller shafts of the other axis.

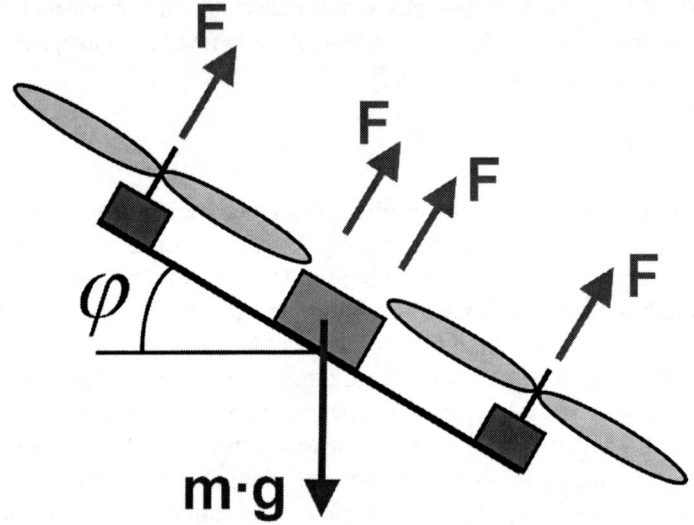

Figure 25: Forces with attack angle φ

The thrust forces are still pointing in the same direction as the motor shaft, and the weight force is still at the center of gravity and points towards the center of the Earth. The forces are no more opposite to each other than was the case for the pure hover.

Forces always act in certain directions; they are so-called vectors. They therefore always have an arrowhead at the end. They can also be decomposed into a sum of other forces. Figure 26 shows how one can understand that. Instead of the force F, two forces can also be drawn. The second starts at the arrowhead of the first one.

F_D is called the drift force, and it acts in the horizontal. The force F_L will be referred to hereinafter as the lift force, and it acts in the vertical. Figure 27 shows an overall arrangement of the decomposed forces. The forces 2 F_L and 2 F_D again belong to the two propellers from the other axis.

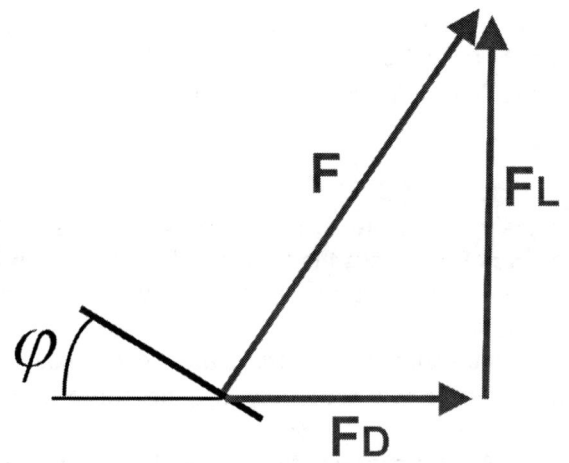

Figure 26: Thrust force F, lift force F_L and drift force F_D

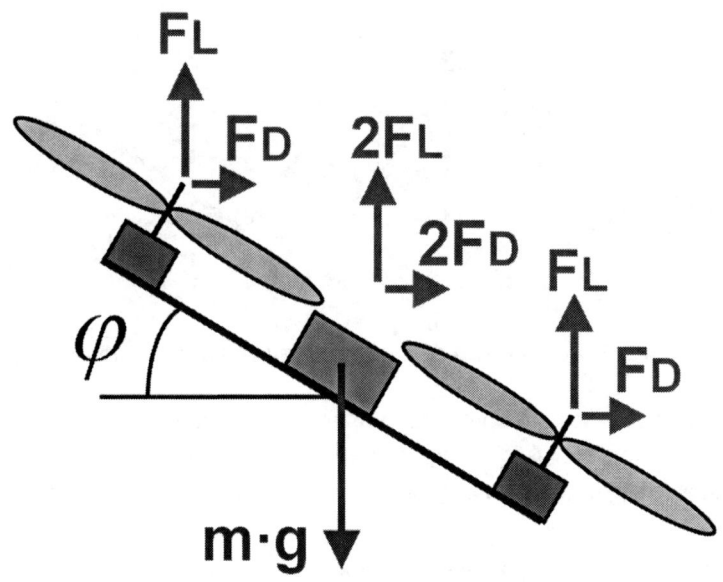

Figure 27: Lift force F_L and drift force F_D

The four lift forces F_L act here again exactly in opposition to the weight force of m · g. To make sure that the Quadrocopter holds its height during drifting, the following equation must of course apply again:

$$4 \cdot F_L = m \cdot g$$

The force triangle in Figure 26 is right-angled. It is easy to see that the thrust force F is always the longest side, the hypotenuse. It is always greater than the lift force F_L and greater than the drift force F_D.

The position of the throttle stick always acts on the thrust force F. If the Quadrocopter is now attacked with φ, according to the above formula, instead of F, only the smaller lift force F_L acts against the weight force. Without changing the throttle stick, the Quadrocopter begins to sink and drifts off to the side (or forward, depending on whether φ acts in the roll or nick direction). The pilot can compensate for this with an increase in throttle position[6].

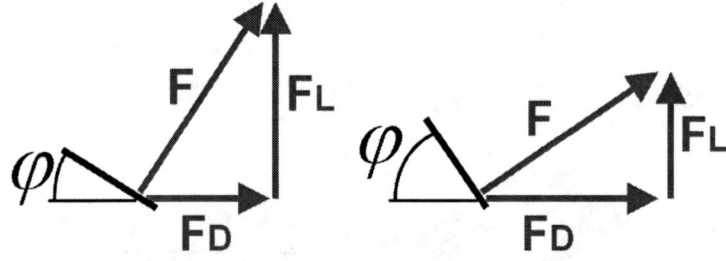

Figure 28: Comparison of two force triangles at different angle φ

Figure 28 shows a comparison as the forces change at a different attack angle φ. It is recognizable that the lift force F_L at a larger attack angle φ and constant thrust F gets smaller in favor of the

---

[6] That's true only if the height controller and the air pressure sensor are turned off. If they are turned on, the force F is set by the controller such that $4 \cdot F_L = m \cdot g$ is always given. For larger φ, the regulator automatically gives more gas to keep the height.

drift force F_D. The Quadrocopter drifts faster in the right-hand illustration, and more gas must be given to keep it hovering.

Once again, 'actio = reactio' should be investigated. This rule is actually violated in Figure 27. To the four vertical forces F_L, the weight force m · g counteracts. But in the horizontal, it seems that the drift force F_D doesn't counteract anything. If one applies a force F to a body in space, so acts F = m · a, Force = mass x acceleration. The body is therefore constantly accelerated by a = F / m. Next is indeed v = a · t, velocity = acceleration x time. If this would match so all, the lateral velocity of the body would increase continuously.

Of course, this does not happen in practice. Indeed, there is something indispensable for Quadrocopters; it is also necessary to ensure that the propellers can generate thrust at all – it is air.

He who in a RC-car presses down the throttle stick for a long time notes that the speed at some point no longer continues to increase. This is exactly the same with a Quadrocopter. The four drift forces F_D are opposed by an aerodynamic drag force F_A.

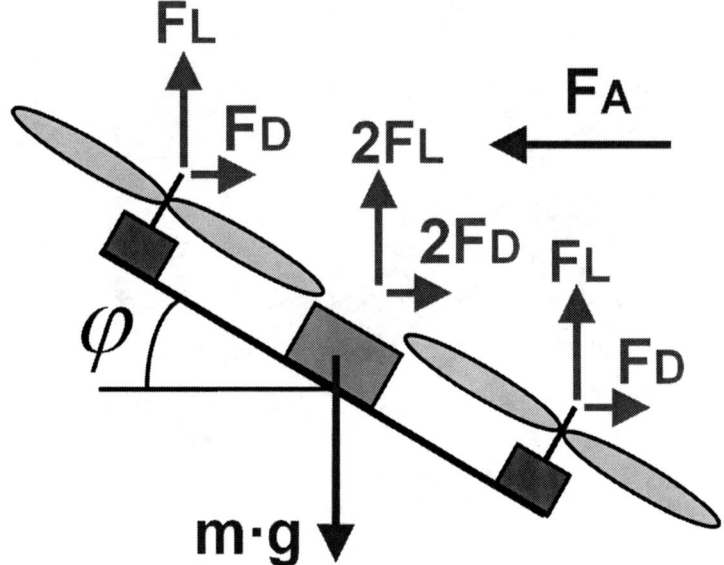

Figure 29: All forces inclusive aerodynamic drag force

This gets larger with increasing speed, and at some point the same size as $4 \cdot F_D$ is reached. Then, the maximum velocity is reached. This is illustrated in Figure 29. The aerodynamic drag force F_A does not attack at a particular point of the Quadrocopter. It acts on the entire surface, so it is drawn 'free-flying'. In the horizontal, it counteracts the drift forces F_D, so at a certain maximum velocity again the rule 'actio = reactio' applies.

4.3 General balance of forces

The mechanisms that allow the Quadrocopter to rotate to the angle φ have not yet been considered. This requires the forces for a general case to be discussed. Without restrictions, it is possible to do this for only two opposing rotors. Figure 30 shows such a general case, in which the force F_1 is not the same size as F_2.

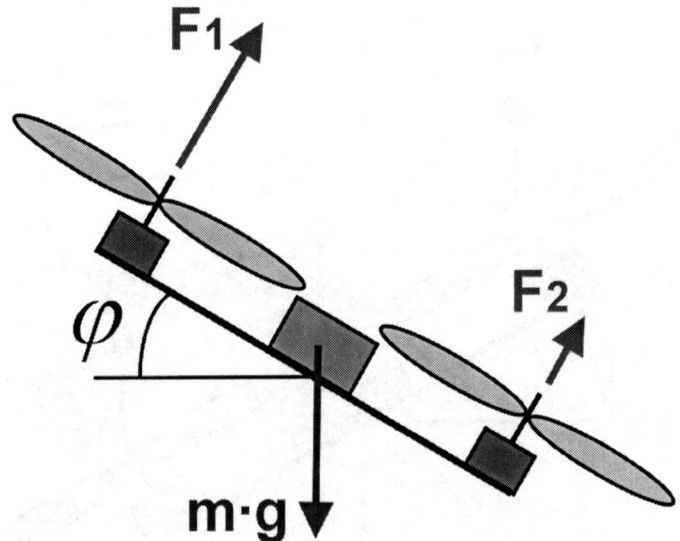

Figure 30: General case: $F_1 \neq F_2$

To imagine what happens in this case, the two forces F_1 and F_2 are each decomposed into two other forces, similarly to the thrust force F being decomposed into the two forces F_L and F_D above. Figure 31 shows this.

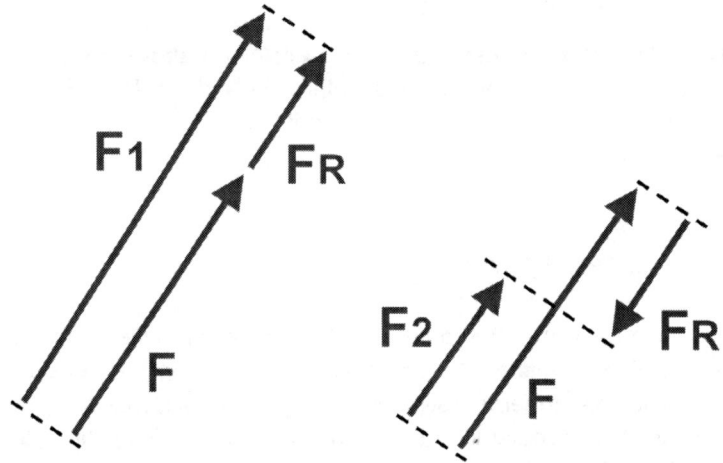

Figure 31: General case, separation of F_1 and F_2 into F and F_R

F_1 and F_2 are decomposed into two forces F and F_R. F is again the previously known thrust force and F_R is a rotational force. Why it is called rotational force is explained below.

It can be seen from Figure 31 that

$$F_1 = F + F_R \quad \text{and} \quad F_2 = F - F_R$$

A numerical example will illustrate this. If F = 5N and F_R = 2N, then F_1 = 5N + 2N = 7N and F_2 = 5N − 2N = 3N. Comparing this with the arrow lengths, this also appears plausible. These formulas can also be transformed so that F and F_R can be calculated from given F_1 and F_2:

$$F = \frac{F_1 + F_2}{2} \quad \text{and} \quad F_R = \frac{F_1 - F_2}{2}$$

The example numbers from above must also comply with these equations: 5N = (7N + 3N) / 2 and 2N = (7N – 3N) / 2

The effect of the thrust force F was explained above. As shown above, it can just be divided into lift force F_L and drift force F_D. F is therefore responsible for 'lift' and 'drift', but cannot cause any rotation of angle φ.

4.4 A simple physical model

It is thus clear that the force F_R – the rotational force – must be responsible for this rotation. In a simplified view only the effect of the F_R will be further investigated (and it would tacitly be assumed that the Quadrocopter still hovers and is possibly drifting due to the not-shown force F).

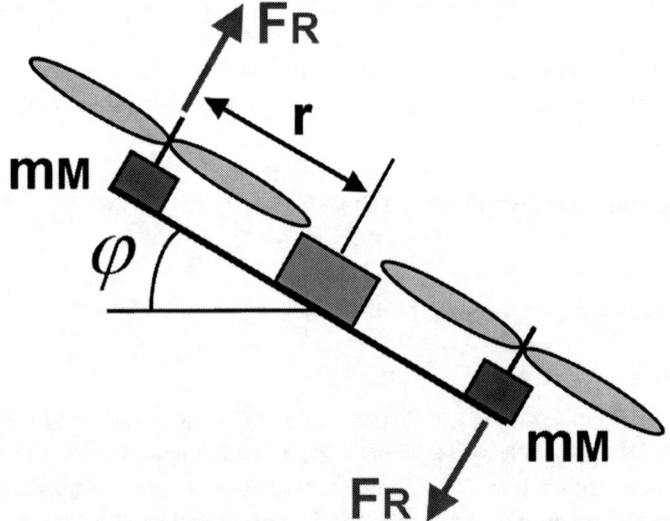

Figure 32: Only F_R is relevant for the rotation around φ

The end of the chapter is again rather complicated. To develop a simple physical model the mass moment of inertia J and the torque M need to be used.

a) $$M = J \cdot \ddot{\varphi}$$

b) $$2 \cdot F_R \cdot r = 2 \cdot m_M \cdot r^2 \cdot \ddot{\varphi}$$

c) $$\ddot{\varphi} = \frac{F_R}{m_M \cdot r}$$

a.) Means that torque = mass moment of inertia x angular acceleration. This formula is not necessarily familiar to everybody, particularly if the physics class was a long time ago. But it can be derived from F = m · a (force = mass x acceleration). The two formulas are in fact similar; while F = m · a is for movement along an axis, this one refers to rotations around an axis. The so-called angular aceleration $\ddot{\varphi}$ is mathematically the second derivative of the angle.

b.) A conversion in the variables that are shown in Figure 32; $M = 2 \cdot F_R \cdot r$ and $J = 2 \cdot m_M \cdot r^2$ apply. In calculating the mass moment of inertia J, it is assumed that the total mass of the axis consists mainly of the two motor masses m_M.

c.) All of that can be dissolved to $\ddot{\varphi}$ and the result is the following mnemonic:

The rotational force F_R acts to the angular acceleration $\ddot{\varphi}$.

It has already been explained in the chapter on the control board that the Quadrocopters balance themselves. The pilot gives only a desired angle φ by the nick or roll stick of the remote control transmitter. This angle is achieved afterwards through internal control. The regulators of this control are discussed in the next chapter, 'Setting the controller'.

The integration of the angular acceleration is the angular velocity and the integration of the angular velocity is the angle itself. In order to calculate the angle φ, the angular acceleration $\ddot{\varphi}$ must therefore be integrated twice. At the very end of the Chapter this is described in a graphical representation. The two integrators can also be presented with a somewhat more theoretical description as 1/s.

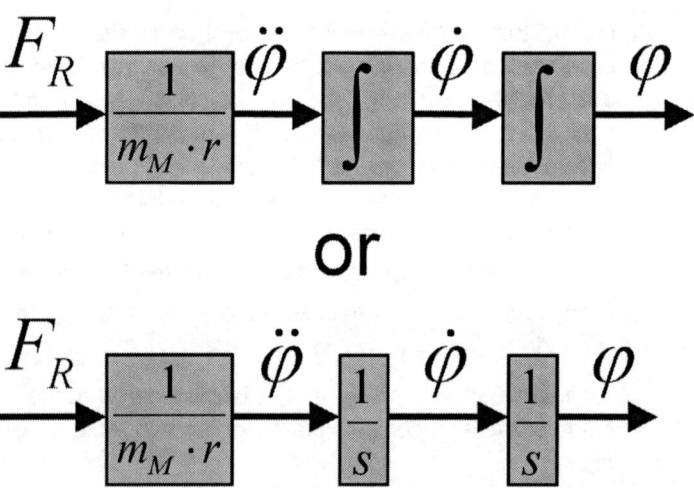

Figure 33: A simple physical model of one axis

4.5 Findings in brief

All too complicated? Here are the key findings of the chapter shown again in brief.

For the pure floating $4 \cdot F = m \cdot g$ applies. The addition of the four thrust forces F acts against the weight force and keeps the model in balance.

If the Quadrocopter now is attacked at φ, according to the above formula, instead of F, only the smaller lift force F_L acts against the weight force: $4 \cdot F_L = m \cdot g$, but it drifts off to the side. The pilot can compensate for this with an increase in throttle position.

To make a rotation through the angle φ possible, the forces F_1 and F_2, generated by two opposing propellers, should not be the same. This results into a rotational force F_R, which is solely responsible for the rotation.

The rotational force F_R acts to the angular acceleration $\ddot{\varphi}$. The integration of the angular acceleration is the angular velocity and the integration of the angular velocity is the angle itself. In order to calculate the angle φ, the angular acceleration $\ddot{\varphi}$ must therefore be integrated twice.

5. Setting the controller

The setting of the controller leads to great discussions in Internet chat forums. The discussions concern the question of how the controller parameters or 'settings' must be chosen so that a certain desired flight performance results. This chapter aims to create a common basis for all systems. Therefore the large number of setting parameters from the various software vendors need to be discussed at the beginning.

On the following pages, only the control of the nick and roll axis is treated, i.e. the previously mentioned 'self-balancing'. The parameters which are responsible for that – K_P and K_D – are always the most important of the whole system, as they are directly responsible for the behavior, such as agility for aerobatics pilots and good-natured for the camera and hovering pilots.

Regulators which contain K_P and K_D are called proportional differrential (PD) controllers. K_P is also called the 'P-part', and K_D the 'D-part'.

5.1 Control of nick and roll axis

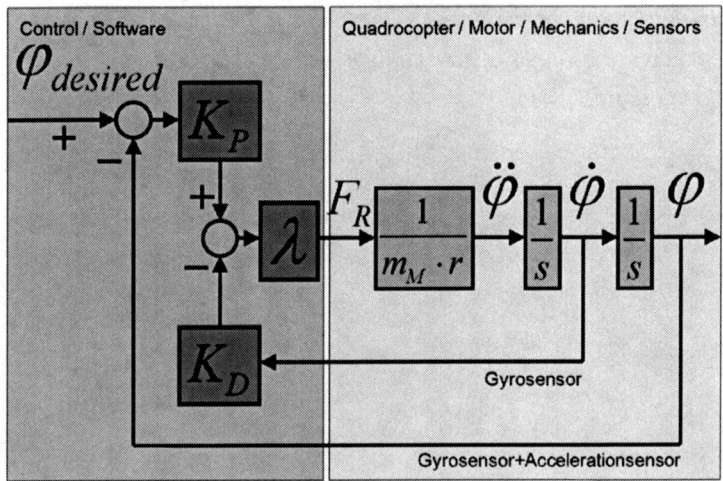

Figure 34: Control of nick and roll axis, block diagram

Figure 34 presents a so-called block diagram of the controlled nick or roll axis. The right side represents the actual hardware of the Quadrocopter. The core forms the physical model of the axis, developed in the last chapter. The left side shows how the parameters of the PD controller are linked to it. It has to be mentioned that the left side is software, and the parameters K_P and K_D also have to be set in the software. The angular velocity $\dot{\varphi}$ is measured by the gyro sensor, and the angle φ with the combination of gyro sensor and acceleration sensor. This was discussed in Section 2.2.

The model pilot commands the desired angle with the nick and roll stick of the remote control. The desired angle φdesired is supposed to be compared with the true angle φ. Therefore the true angle is subtracted from the desired angle and the result is multiplied by a factor K_P (for K proportional). K_P will have the task of influencing the rotational force F_R so that the desired angle and the true angle are the same after a while.

The signal of the gyro is the time derivative of the angle. This signal is amplified by a factor K_D (for K differential). Afterwards, it is also acting onto the force F_R. One could prove that the angle could not be stabilized without the additional gyro signal and the factor K_D. This will be overlooked for now.

There still remains the question of the last factor, λ. This was only introduced in order to keep things general, because electronics and software vendors use different scales. To create a force, a speed reference has to be sent to a brushless controller via an interface. This is done digitally, so on the 'bits and bytes level'. It is scaled differently in the systems. It is therefore not surprising that the parameters K_P and K_D are around 10 with one manufacturer, around 100 with another and around 1000 with the next. Sometimes different names for K_P and K_D are also used. With microcopter for example, K_P is called 'Gyro-I' and K_D 'Gyro-P'. But we should not allow ourselves to get confused about this.

5.2 Effect of K_P and K_D

Thus, it is now possible to describe exactly how different values of K_P and K_D effect the behavior of the Quadrocopter. This will not be carried out 'by hand', but with a simulation program. There, the block diagram of Figure 34 is simply entered and the curves of the angle can then be easily displayed.

The simulations are made for a standard Quadrocopter running with the control electronics. The data are: 60g motors, maximum thrust 2400g, 800g weight, 45cm spacing of axes. It is always assumed that, at the beginning, the system is hovering and the pilot suddenly commands a desired nick angle of 10° via the remote control.

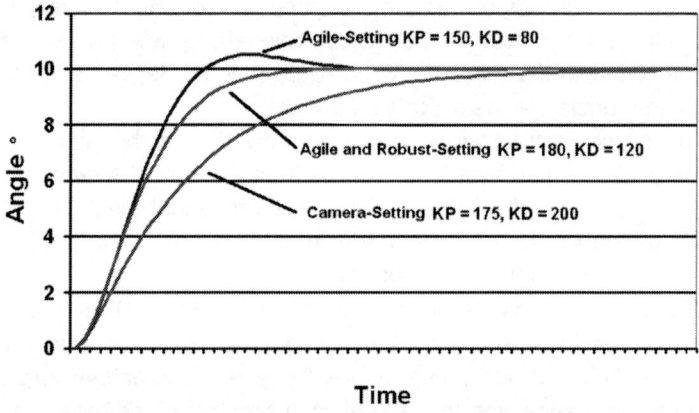

Figure 35: Angle curve φ for different values of K_P and K_D

Figure 35 shows some angle curves for different settings. The camera setting displays a very good-natured behavior. The angle is slowly approaching 10°. If the pilot makes a steering error, this only has an effect on the Quadrocopter after a time delay. Therefore counter-steering is still easily undertaken.

A comparison of the 'Camera Setting' with 'Agile Setting' at 8°, i.e. when almost the full scale of 10° is achieved, shows something interesting. Using 'Agile Setting', the controller requires only about half as much time to reach this angle. This also means that steering mistakes become apparent here much more quickly.

This 'Agile Setting' has yet another feature: a so-called overshoot. At one point, the angle is slightly greater than 10°, before it then falls back to that value. This is normal. Systems in which the control parameters are set too 'fast' are always firing something beyond the target.

A model pilot can also see these characteristics of the settings in his aircraft. For that purpose he pushes the stick of the remote control back and forth while the Quadrocopter is hovering.

Using 'Camera Setting' it can be observed that the Quadrocopter 'is not hanging on the stick', but that the achieved angle always lags something behind to the desired angle. This is shown in the graph in Figure 35. The angle of 10° desired by the pilot is achieved far more slowly than in other settings.

This is of course nothing for a speed flying pilot, but ideal for camera or hover flights.

Using 'Agile Setting', the Quadrocopter 'hangs on the stick' more. An ideal 'hanging on the stick' would be that all movements of the remote control's stick can be observed immediately at the aircraft. This is not possible because the drives can't change the speed fast enough, the thrust forces are limited, the transmission of the remote control is too slow, etc. The speed and aerobatic pilot is nonetheless happy, because everything he does on the remote control quickly arrives at the model. However, the camera pilot doesn't like that setting at all, because steering errors of course also arrive quickly.

In Figure 35, a third setting is illustrated, which is called 'Agile and Robust Setting'. It is a compromise between the other two. It is on the one hand almost as agile as the 'Agile Setting' but has, on the other hand, no overshoot. Of course, the author has also tested this setting on his Quadrocopter, and found it as stable as piece of wood in the air that can still act as agile as a Ferrari.

Squirrelly and fidgety

But it is not only the steering mistakes which become noticeable more quickly for the one setting and slower for the other. Exactly the same applies to interference from wind or noise signals from the sensors. Therefore, a 'Camera Setting' Quadrocopter is smoother in the air, also in a hover flight, with no additional commands.

This fact also helps the camera pilot. To keep the camera still when photographing is ultimately the most important measure against blurred images. In contrast, a system in the 'Agile Setting' behaves much more squirrelly and fidgety. It immediately tries to steer a disturbing influence, and furthermore it overshoots.

Changing K_D while keeping K_P the same

Anyone who has ever tried to adjust the K_P and K_D with a PD controller for a desired behavior has realized that this is not such a simple task. There are just two parameters which are independent, and therefore there are almost endless possibilities. To investigate the influences systematically, the following procedure is pursued in Figures 36 and 37.

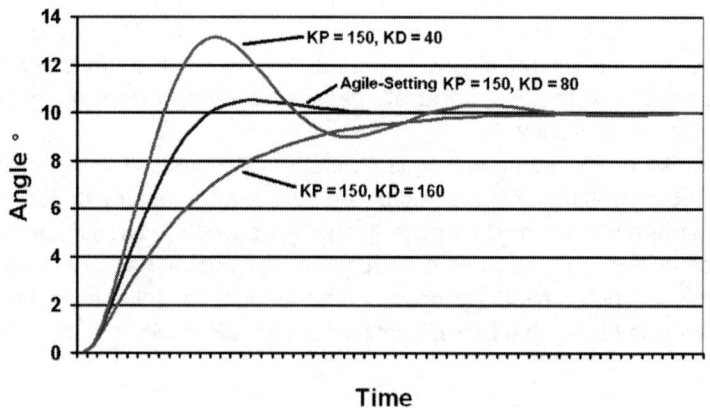

Figure 36: Angle curve φ by changing K_D

It is always started with the 'Agile Setting'. Then, one of the two parameters remains the same and the other is changed.

Figure 36 shows a change of K_D, while K_P always remains at 150. One can clearly see that the overshoot is very large with too small a K_D, while the system is getting even more agile. With those settings, the Quadrocopter would actually be almost un-flyable. An increase in the K_D, however, makes the system good-natured.

If K_D remains at 80 and K_P is changed as shown in Figure 37, the system behaves with a K_P of 200 a little more agile and for a smaller K_P good-natured again.

Generally the following observations can be made:

For smaller K_P and larger K_D, the system is good-natured, slow and less susceptible to interference.
For larger K_P and smaller K_D, the system is more agile and responds quickly to disturbances.
If K_D is too small, the system is less damped and therefore tends to overshoot.

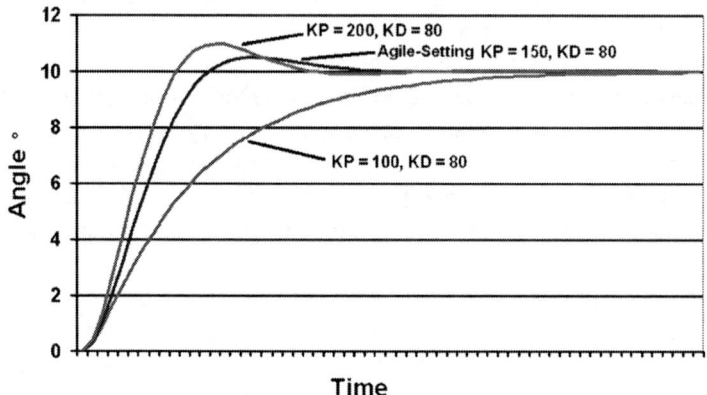

Figure 37: Angle curve φ by changing K_P

Best practice by setting the controller

If the Quadrocopter is built with the motors, propellers and spacing of axes as recommended by the manufacturer, a suggested good-natured beginner setting would allow a stable levitation. The further offered settings – 'Sport', 'Agile', 'Advanced', 'Expert', 'Camera', 'Beginner', and the rest of them – often provide sufficient elbowroom for most users and they can be easily adopted. If a Quadrocopter doesn't hover with the suggested components and settings, it is often not the K_P and K_D parameters but other aspects that are responsible. In such a case reference is made to the chapter 'Initial operation, sources of error and first flight'.

Things get more interesting when other motors, propellers and spacing of axes are used.

Figures 36 and 37 show very nicely how the overall behavior is changed with varying K_P and K_D. In practice, however, the model pilot doesn't have such graphs for 'his' aircraft on hand. He sets the parameters with the PC interface, starts the motors and can only find out visually whether it behaves as desired. So he must 'aviate' the parameters.

Therefore he starts with a good-natured 'beginner setting'. If the Quadrocopter hovers, the optimization can be started, depending on the desired behavior.

Agile behavior: Decrease K_D or enlarge K_P, until an overshoot is visible. You see that at first the angle deflects further than at the end. If the system already overshoots strongly while running the startup parameters, then enlarge K_D a little and decrease K_P.

Camera behavior: Enlarge K_D or decrease K_P until the Quadrocopter is visibly lagging behind, i.e. no longer 'hanging on the stick'.

Another approach may be: starting from a 'Beginner Setting', enlarge the value of K_P and decrease K_D by the same factor to effect more agility, or decrease the value of K_P and increase the K_D by the same factor to cause more good-natured behavior.

5.3 Transfer function

To learn something more about the influence of parameters, and to provide even more setting tips, will once again be introduced to a somewhat theoretical, but as the subsequent examples show, very practical subchapter. From the block diagram in Figure 34, a so-called transfer function can be calculated. It is equivalent to the ratio between the true angle φ and the desired angle φdesired. It reads:

$$\frac{\varphi}{\varphi_{desired}} = \frac{K_P \cdot \frac{\lambda}{m_M \cdot r}}{s^2 + s \cdot K_D \cdot \frac{\lambda}{m_M \cdot r} + K_P \cdot \frac{\lambda}{m_M \cdot r}}$$

Figure 38: Transfer function

The 's', which can be found in the denominator, may allow some dynamic studies to be made using this transfer function. The factor 'λ' was discussed in Section 5.1. 'm_M' and 'r' are provided in Figure 32. For the considerations at this point, special attention is to be directed to only a single key issue: The temporal variation of the angle and thus the behavior of the Quadrocopter remains exactly the same if all terms of the equation also remain the same[7].

The example of the Quadrocopter calculated above is again considered: 60g motors, 2400g maximum thrust, 800g weight, 45cm spacing of axes. For this an 'Agile Setting' with a K_P of 150 and a K_D of 80 was found.

---

[7] Terms for the present example: $K_P \cdot \frac{\lambda}{m_M \cdot r}$ and $K_D \cdot \frac{\lambda}{m_M \cdot r}$

If you now want to change things structurally and find new values for K_P and K_D so that the behavior remains the same as before, you must only ensure that

$$K_P \cdot \frac{\lambda}{m_M \cdot r} \quad \text{and} \quad K_D \cdot \frac{\lambda}{m_M \cdot r}$$

remain the same.

1st example: *The spacing of the axes will be increased from 45 cm to 60 cm. Otherwise everything remains the same. r is in the denominator in both terms and therefore changes them to 1 / (60/45), so both K_P and K_D must be greater by 60/45. The new values are K_P = 150 x 60/45 = 200 and K_D = 80 x 60/45 = 107.*

2nd example: *Smaller motors of 30g are installed. Thus, this changes both terms by 1/(30/60). K_P and K_D would have to be reduced by half. But in addition, the two drives generate only about half the thrust[8]. This means that at a certain speed only about half as much thrust is available. Thus, λ is also approximately reduced by half. K_P and K_D would therefore have to be doubled to compensate again. The new values are K_P = 0.5 x 2 x 150 = 150 and K_D = 0.5 x 2 x 80 = 80. The values can therefore be left unchanged.*

Example 2 is also the reason why so many pilots realize a similar axis distance of maybe 35cm to 45cm and fly with the same controller settings with comparable properties, even though they are installing different motors and propellers with different power.
Therefore, in various Internet forums, parameter settings are compared and discussed, even though many different motors and propellers are used. Because a heavier drive also provides more

---

[8] This is only rudimentarily true, since the relationship between speed and thrust is not linear. But Chapter 6 ('Dimensioning of motors and propellers) shows more.

thrust, the influence on the control parameters K_P and K_D is minimal.

5.4 Heading hold

Until now, it was assumed that a Quadrocopter must be equipped at least with gyros and accelerometers to ensure that it can be balanced by the controller. This must be refined somewhat at this point.
The very first Quadrocopters contained no acceleration sensors at all, but only gyros. There are also several systems that are equipped in this way to this day.
This means that there is no more control of the absolute angle, but a control of the angular velocities. This happens in all three angular degrees of freedom – nick, roll and yaw.
A certain stick position on the remote control corresponds to a desired angular velocity. The center position corresponds to zero angular velocity. Zero angular velocity means that the angle reached at this time is kept apart from the sensor drift. Hence the name 'heading hold' or 'angle hold'.

For aerobatic and speed pilots

It was already discussed earlier: The more degrees of freedom are controlled by the system itself, the fewer degrees of freedom the pilot can control. This can be boring for him. In the heading hold mode, even fewer degrees of freedom, around 'half' of the angular degrees of freedom, are regulated, so more is demanded of the pilot.
Many have therefore to this day remained loyal to the heading hold mode, because they say the controlling is more challenging and the flight more exciting. Especially for the speed and aerobatic pilots that is very often the case.
The attack of an angle with a short nick motion can be compared with the pitching of the elevator on a model airplane. When the

stick there is pushed back again to the neutral position, the aircraft also holds the angle. Anyone who has previously flown model aircraft will possibly therefore prefer heading hold for forward flying. But everybody has to try it out for themselves. More heading-hold-like flight behavior is achieved by decreasing K_P. However, one has to be careful, because not every system allows a very small K_P.

Loops

Loops must necessarily be flown in heading hold mode, as the nick or roll angle must briefly be greater than a maximum value.
Some systems can be configured to switch briefly into the heading hold mode with a fast motion of the nick or roll stick. In other systems it is switched by a channel of the remote control. The flying of the loops requires a little practice and, especially at first, enough air under the propeller.
A strong move of the nick stick should be accompanied by a gas removal. Ideally, the speed during the stroke is very low because the thrust is then pointing towards Earth. In, or shortly after the rollover, the gas is increased again and the nick stick is in the neutral position. Thus, the Quadrocopter will hopefully be in balance again.

Comparison with model helicopters

In the yaw axis, model helicopters have only a gyro for control. In the nick and roll axis there are passive stabilizers. A stick movement acts on the swash plate; the helicopter tends with increasing time more and more in the direction of the stick. So the stick acts also here on the angular velocity and not on the angle.
The control of a helicopter can therefore most likely be compared with the heading hold mode of the Quadrocopter.
Beginners often ask what is now more difficult to steer. There are many partly contradictory statements on this subject, often be-

cause of personal experiences. From a technical point of view we can say that a Quadrocopter in heading hold mode is as difficult to control as a helicopter. A Quadrocopter equipped with the base components as discussed in Chapter 2 is significantly easier to steer. It has gyros and accelerometers, so the angle is therefore controlled in nick and roll.

If we assume that it is easier to steer a flying machine with an increasing number of controlled degrees of freedom, then:

Helicopter, 3 x '0 .5' degrees of freedom controlled	difficult
Quadrocopter in heading hold: 3 x '0 .5' degrees of freedom controlled	difficult
Quadrocopter with standard components: 2 + '0 .5' degrees of freedom controlled	easy
Quadrocopter with standard and expansion components, fully expanded: 6 degrees of freedom controlled, 'air nail':	Pilot has no mission

It is also true that the helicopter pilot is able to immediately control a Quadrocopter. He may only wonder how smooth and quiet-running this is in comparison. Pitch-controlled systems, where the rotors are screwed together, have much larger problems with unbalance. Speed-controlled propellers, which are made of one piece, are much less problematic here. In the opposite case, expressed in general, for a Quadrocopter pilot it is more difficult to control a helicopter. Since both systems can be controlled with the same functions, however, it is still relatively simple to switch from one to the other.

5.5 Findings in brief

As in the last chapter, here the main findings are presented in abbreviated form.

The parameters K_P and K_D are always the most important of the whole system. They are used for the balancing of the nick and roll axis.
The producers of Quadrocopters often use different names for those parameters.

With smaller K_P and larger K_D, the system is good-natured, slow and less susceptible to interference. With larger K_P and smaller K_D, the system is more agile and responds quickly to disturbances.

If K_D is too small, the system is less damped and therefore tends to overshoot.

An easy approach for parameter adjustment may be: Starting from a 'Beginner Setting', to enlarge the value of K_P and decrease K_D by the same factor to effect more agility, or decrease the value of K_P and enlarge the K_D to the same factor to cause more good-natured behavior.

Because a heavier drive also provides more thrust, the influence on the control parameters K_P and K_D is minimal.

In the heading hold mode, only the angular velocities (instead of angles) are controlled. This type of control can be compared most closely with that of a helicopter.

6. Dimensioning of motors and propellers

What are the tires on a Formula 1 racing car are the propellers on Quadrocopters. The tire brings the huge torque of the engine onto the asphalt. This creates a driving force and hence an acceleration. The propeller generates thrust through the air. As described in Chapter 4, this is also responsible for the movement.
No-one would come onto the idea of installing a steam engine in the Formula 1 car and using it to drive the wide, high-tech tires. On the other hand, surely no-one would think of mounting an old wagon wheel onto racing cars. It is vital that engine, tires and gear must be coordinated.
Exactly the same is true for the flight model. To obtain a satisfactory result, the right motor and propeller combination must be selected. Also in this section some theory is considered. It is intended as a background and continuation of the two already mentioned pieces of information:

For motors with a kV of about 1000 rpm/V:
For a motor with 30g weight, an 8" x 4" propeller in direct drive is selected, which gives about 4 x 300g maximum thrust at 10V.
For a motor with 60g weight, a 10" x 4" propeller in direct drive is selected, which gives about 4 x 600g maximum thrust at 10V.

6.1 Propellers

Always make clear that the maximum allowed speed of the used propeller is higher than the speed of the used motor. Some basic terms used later need to be explained first.
Propeller diameter D: the diameter of the propeller, measured from blade tip to blade tip
Propeller pitch H: Since the propellers are twisted, the manufacturers often define the pitch differently. As a rule of thumb, take the pitch of the propeller at $0.7 \cdot (D / 2)$, measured from the center. Then you imagine that with this angle one cuts butter with a

complete revolution. The distance covered during the revolution is the propeller pitch H.

Propeller area F: the area which envelops the rotating propeller: $F = Pi / 4 \cdot D^2$

The power of a moving aircraft is equal to the thrust x speed of flight, v. the formula is $P = F \cdot v$. Fast aerobatic and speed models must still be able to develop at a speed of perhaps 150 km/h thrust for the flight figures.

On the other hand there are the slow and park flyers. These fly very slowly in comparison. In indoor flights, you often see them 'torqueing'. This is when they execute a rolling movement in one place. Here, the model is vertical in the air. Then it has no aerodynamic lift, but just the thrust of the propeller. The model doesn't move; the flight speed is zero. The thrust is in this case also called static thrust force. Strictly speaking, the power converted in terms of the model will also be zero (static thrust force x 0 = 0).

The Quadrocopters are therefore more comparable to the slow and park flyers. In their case, the hovering flight is a common situation. It is clear that the propeller must be designed in different ways for different applications.

Propellers for aerobatic and speed flight models usually have a smaller diameter D at a greater pitch H, compared to the slow-flyer propellers with the same power. To explain this, one can make use of the stream theory. There, it can be assumed that the propeller sucks the air equally on the surface of the propeller area.

In order that the speed model still generates thrust at adopted 150 km/h, the air jet must obviously be faster than the model itself. This requires a large propeller pitch H. A slow flyer or Quadrocopter doesn't need a fast jet speed; as it is often operated in hover, the pitch is small. For the same power, the propeller diameter D, therefore, can be made slightly larger. As rules of thumb we can say that:

Aerobatic or speed models: pitch to diameter ratio $H / D = 0.7 ... 1.0$

Slow flyer or Quadrocopter: pitch to diameter ratio H / D = 0.3 ... 0.7

In previous chapters 8" x 4" or 10" x 4" were given as a rule of thumb as suitable propellers for Quadrocopters. With H / D = 0.5 or 0.4, respectively, these very clearly belong in the 'slow flyer' or 'Quadrocopter' divisions. They are designed for static thrust and not speed.

Specific thrust

For static thrust force and required power, the stream theory provides formulas. Only the results are presented here. More detailed calculations can be found in the literature (1).

$$S = \sqrt[3]{2 \cdot \rho \cdot F \cdot (\xi \cdot \eta \cdot P)^2} \quad \text{and} \quad P = \frac{1}{\xi \cdot \eta} \cdot \sqrt{\frac{S^3}{2 \cdot \rho \cdot F}}$$

ξ: Propeller efficiency, 0.5 is the value for an average model aircraft propeller
η: Motor efficiency; here 0.7 is assumed
ρ: Density of air, 1.24 kg/m³
F: Propeller area, as defined above, F = Pi / 4 · D²

In Section 2.5, in the dimensioning of LiPo accumulators, a value of 100W of electrical power required per 1kg levitated mass was calculated. That is to say that every one watt of power transports 10g mass in the air. The so-called specific thrust is calculated as 10g/W or, more scientifically, with the unit Newton[9]: S/P = 0.1N/W. This value will now be investigated in more detail.

1st example: Using the above formulas, it can be calculated how large the specific thrust really is. Several systems are calculated

---

[9] The unit problem 'grams' and 'Newton' was discussed in Section 4.1.

with 8", 10" or 12" propellers. As an example, a 800g Quadrocopter is taken. For an 8" propeller it would be a bit tight; you remember that a maximum of about 4x300g thrust is produced, but this could still float. S must be expressed in Newton; 800g corresponds to 8N. Since a total of four propellers are involved together, the area is a total of $4 \cdot Pi/4 \cdot D^2$. D must be measured in meters. An 8" propeller has a diameter of $8 \cdot 0.0254m = 0.2032m$. Using a 10" propeller, the value is 0.254m and using a 12" propeller 0.3048m.

The calculation with the formula for the performance revealed something interesting:
8" propeller: P = 113W; S/P = 8N/113W = 0.071 N/W or 7.1 g/W
10" propeller: P = 90.5W; S/P = 8N/90.5W = 0.088 N/W or 8.8 g/W
12" propeller: P = 75,4W; S/P = 8N/75.4W = 0.106 N/W or 10.6 g/W

The previously assumed value of 10g/W is not far away from these numbers. Obviously, the larger the propellers are, the better the Quadrocopter is energetically. Four 8" propellers still need 113W to lift the 800g. The 12" propellers need only 75W. In practice, 12" propellers are also driven with larger motors. So they are heavier. The higher specific thrust means that you need for a heavier Quadrocopter with larger propellers a similarly large battery as for a smaller one.
Of course, this is true for the entire flight (model) construction. It is also responsible for ensuring that a helicopter has very large rotor blades. At this point we may pause for a moment.
'Are Quadrocopters which make use of small 8" to 12" small propellers a bad design in comparison with model helicopters, concerning the specific thrust?' one might ask. 'In principle, yes!' is really the answer.
Since, however, the available brushless outrunner motors are designed for direct drives of such propellers, the mechanics are simple and a gearbox is also unnecessary. So the effectivity remains high and one can achieve flight times of 10 to 30 minutes

with normally big accumulators. Therefore, these are comparable with helicopters.

6.2 Larger propellers with gearbox

In order to realize a Quadrocopter with greater specific thrust, here the complete design for the use of 20" propellers of a coaxial helicopter is presented[10].
The diameter D is 0.512m. As in the example, the needed power and the specific thrust for a 800g model can be calculated as

20" propeller: P = 45.2W; S/P = 8N/45.2W = 0.177 N/W or 17.7 g/W

Figure 39: 60g motor, gear 16/50, 50 cm ø propeller of a coax heli

---

[10] Coaxial helicopters: These have no tail rotor, but two left- and right-rotating main rotors on one shaft. A yaw movement is realized with different rotor speeds. Many smaller model helicopters are designed as coaxial helicopters.

Compared with the 8" propeller, the required power is smaller by a factor of 2.5! This is almost incredible! To drive the propeller, we would now like to use a normal brushless outrunner motor. Therefore, a gearbox will be dimensioned.

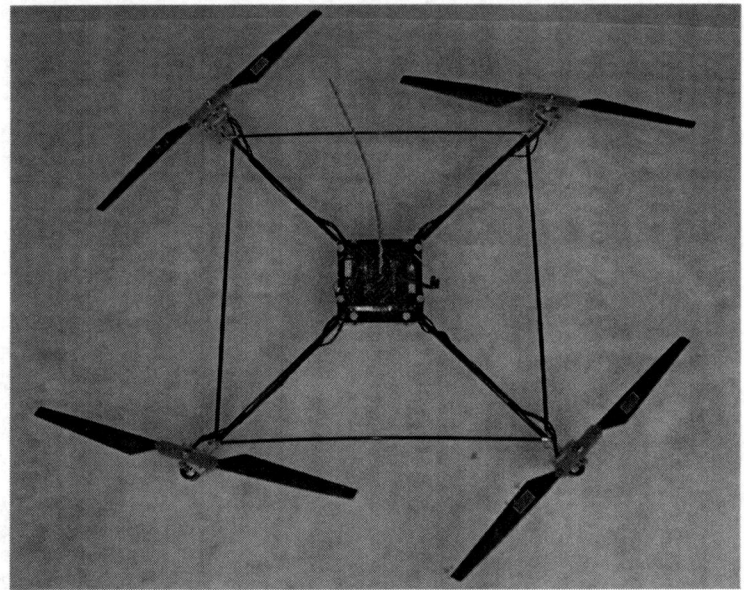

Figure 40: Quadrocopter with 80 cm spacing of axes and 20" rotors

A typical gearbox design would be calculated by the desired maximum power at a certain maximum speed. One could then compare this with the maximum speed of a standard combination 60g motor and 10" propeller and determine the gear ratio accordingly. This method is only of limited significance, because something else much more important must be considered: the mass moment of inertia of the propeller.

Since the control of nick and roll angle is realized over the motor speed (see Chapter 5), it is important to ensure that the motor gets through the gearbox the same (rather smaller, certainly not bigger) mass moment of inertia as in the standard configuration 60g motor and 10" propeller. Only in this way are the dynamics of motor and propeller combination fast enough.

The mass moment of inertia of a propeller is calculated greatly simplified as that of a rod by:

$$J = \frac{1}{12} \cdot m \cdot D^2$$

A standard 10" (D = 0.256m) propeller has a weighted mass of 17g. The weight of the 20" (D = 0.512m) propeller with the rotor blades of a coaxial helicopter is 40g. The two moments of inertia are calculated as $93 \cdot 10^{-6}$ kgm² and $874 \cdot 10^{-6}$ kgm². The moment of inertia of the 20" propeller is thus about 9.4 times greater. Using a gearbox, the moment of inertia is not translated linearly, but with the square of the reduction. In order that the motor therefore gets the same moment of inertia with the 20" propeller, the square of the reduction must be equal to 9.4. This results in a transmission of at least 3.07:1 (3.07 is the square root of 9.4).

This Quadrocopter is a little heavier because of the gearbox, but it hovers for longer with the same battery than one with 8" or 10" propellers.

6.3 Motor

In all formulas of the last section the motor speed – i.e. the size, 'revolutions per minute' ('rpm') with the abbreviation 'n' – is missing.

n can also be calculated if one works with special coefficients. The most important are the thrust coefficient C_T and the power coefficient C_P. These values are also linked to the already previously introduced propeller efficiency 'ξ', namely via the following formula from source (1):

$$\xi = \sqrt{\frac{2}{\pi}} \cdot \sqrt{\frac{C_T^3}{C_P^2}}$$

Via propeller efficiencies and coefficients a series of measurements and other theoretical considerations can be carried out. These are also described in the source (1). For the further calculations with a typical slow-flyer or Quadrocopter, propellers in sizes from 8" to 12" and the pitch to diameter ratio H / D from 0.4 to 0.5, the following numbers are applied:

$$\xi = 0.5;\ C_T = 0.085\ \text{and}\ C_P = 0.04$$

Using these values, the static thrust and the power can be calculated under consideration of the speed. The formulas from source (1) are made for a single propeller.

$$S = C_T \cdot \rho \cdot \left(\frac{n}{60}\right)^2 \cdot D^4 \quad \text{and}$$

$$P = C_P \cdot \rho \cdot \left(\frac{n}{60}\right)^3 \cdot D^5$$

2nd example: *For a Quadrocopter with 10" x 4" propellers and a speed of 4000 rpm, the static thrust and the power are calculated. If one sets D = 0.256m, ρ = 1.24 kg/m³ , C_T = 0.085 and C_P = 0.04, the total static thrust is 4 x 2.01N = 8.04N. So with this speed, about 800g can be lifted. The power P is in the above formula 4 x 16.15W = 64.6W. It is, however, the propeller power. If you want to calculate the total power of the motors, this value must be divided by the motor efficiency η. If this is assumed to be 0.7, as in the 1st example, the result is 92W. This value is slightly higher than there (90.5W), but the static thrust is also slightly higher because of the use of a 'round' speed. Moreover, the C_T and C_P values are also slightly rounded. The current that flows in this case is calculated for a three-cell LiPo accumulator as 92W / 11.1V = 8.3A.*

3rd example: *For the same Quadrocopter now the maximum static thrust can be calculated. Assuming that the drive makes 1000 rpm/V, and is operated with a maximum of 10V, then the maximum* speed is 10,000 rpm. Using these numbers so, a maximum thrust of 50N or 5000g is obtained with all four drives together. This value is about twice as high as in the rules of thumb indicated above, where there is only 4 x 600g = 2400g. In fact, the drive produces at the applied 10V not 10,000 rpm, but only around 6000 to 7000 because of the high current and, in some systems, because of safety restrictions by the software. If the value of 7000 rpm is used in the formulas, a maximum thrust of 4 x = 6.16N = 24.64N, or 2464g, is obtained. The electrical power of motors (with an efficiency of 0.7) becomes 4 x 123.5W = 494W. The maximum current is calculated for a three-cell LiPo accumulator as 494W / 11.1V = 44.5A.

For the static thrust and power formulas an interesting mnemonic can still be found:

The static thrust varies with the square of the speed, whereas the required power varies with the cube of speed.

In other words, the higher the speed, the smaller the specific thrust, because the power will grow stronger than the thrust.

4th **example:** For the 2nd and 3rd examples (same motor and propeller, operated at 4000 rpm or 7000 rpm), in each case the specific thrust can be calculated.
For the 2nd example S/P = 8.04N / 92W = 0.087 N/W or 8.7 g/W
For the 3rd example S/P = 24.64N / 494W = 0.050 N/W or 5 g/W
The mnemonic is therefore confirmed.

The motor mass is roughly a measure for the power. Motors of 30g weight should be for a maximum output of not more than about 75W. 60g motors are designed for a maximum of about 150W. For the above examples, therefore, 60g motors should be used. However, because often only the 'floating power' of the 2nd example is required, they are greatly oversized with 92W / 4 = 23W continuous power. On the one hand, the remaining is reserve, if you think of the earlier-discussed thrust to weight ratio of 3:1. On the other hand, the motors are mostly complete in the propeller air stream due to the mounting. Quite in contrast to the use in model aircraft in which they are often run with maximum power, so here they are in flight usually only lukewarm.

6.4 Three- or four-blade propellers

For the use of three- or four-blade propellers, source (1) also provides the mnemonics:

For a three-blade propeller, the thrust increases at the same speed by a factor of 1.4 and the power by a factor of 1.6, compared with the two-blade propeller of the same design.
For a four-blade propeller, the thrust increases at the same speed by a factor of 1.8 and the power by a factor of 2.2, compared with the two-blade propeller of the same design.

There are very few Quadrocopters which are not driven by a two-blade propeller. The reason is as follows. Whether you use three- or four-blade propellers, the power output is always growing faster (by 1.6 and 2.2) than the generated thrust (by 1.4 and 1.8). Such propellers are therefore always less efficient than two-blade propellers, if the same diameter is assumed.

6.5 Power and thrust measurement

Once the Quadrocopter assembled we would like to know how big the static thrust and the power output actually are. It is possible to take both measurements using a voltmeter and a normal kitchen scale.
In the previous chapters the total power of the drives was always calculated. Since the consumption of the control electronics is small compared to the drives, the power can be measured directly via the battery. The voltage measurement is not a problem here, but rather the currents, which are in the range between 5A and

50A. They can't be measured directly with commercially available multi-meters, as these are seldom designed for more than 10A.

A shunt resistor of 0.01Ω in series between accumulator and control electronics is the solution. At a current of 20A, $20A \cdot 0.01\Omega = 0.2V$ is then measured over its connectors. The voltage drop is minimal compared to the around 10V of the battery and only slightly distorts the measurement. The shunt should be designed for even a few watts otherwise it burns out at high currents. A 3S LiPo accumulator then supplies at 20A current the power $11.1V \cdot 20A = 222W$.

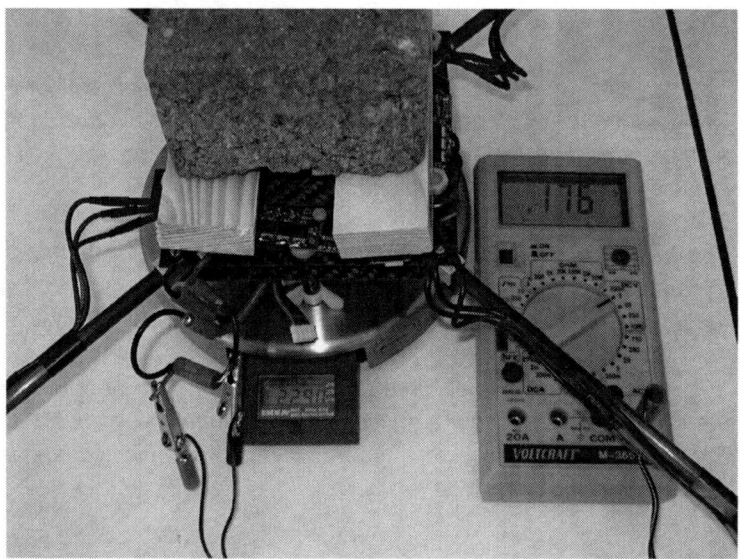

Figure 41: Thrust measurement using a kitchen scale, current measurement with a shunt resistor. Here the current is: $0.176V / 0.01\Omega = 17.6A$.

Since the middle part of the Quadrocopter doesn't generate thrust, it can be easily placed on a kitchen scale. It should then be somewhat weighted with stones, for example. Care must be taken that no propeller blows directly onto the kitchen scale, otherwise the results will be distorted. The difference in weight between full throttle and then being stationary is the maximum thrust. If possible, a distance equivalent to the diameter of a propeller should be kept to the ground, as otherwise the measured thrust is a little too high because of the so-called ground effect.

6.5   Findings in brief

As in the previous chapters here follow the main findings in brief.

For Quadrocopters the same propellers are used as for slow flyers. The pitch to diameter ratio H/D is usually between 0.4 and 0.5.

The specific thrust increases with a larger propeller diameter. Quadrocopters with small 8" or 10" propellers therefore only achieve attractive flight times because they can be operated with brushless outrunner motors in direct drive.

The faster the propeller rotates, the smaller is the specific thrust, because the power increases with the third power of the speed, but the thrust only with the square.

30g motors should be designed for a maximum power of not more than 75W, 60g motors for a maximum power not exceeding 150W. For motors with other weights, the maximum performance changes approximately in proportion.

7. Special shapes, Tri-, Hexa-, Octocopters

The majority of flying multi-rotor systems are equipped with four propellers. But there are special types of constructions with fewer or more rotors. They all have different advantages and disadvantages compared to the Quadrocopter. Discussing these is the subject of the present chapter. At the end, a few pictures of flying Depron bodies will be shown.

7.1 Tricopters

If one were to ask how many speed-controlled propellers are actually needed to keep a multi-rotor system stable in the air, the answer would be three.

Figure 42: Tricopter with camera mounting, Photo: Karl-Heinz Wolf

When using an odd number of propellers however, the same number of left- and right-rotating propellers can't be used. Thus, the yaw torque must be compensated in a different way. The Tricopter constructors usually use from the beginning only right-turning propellers. Since these belong to the standard range in model construction, they are present in much greater choice in the market. One is then not forced to choose a propeller that has a right- and left-turning design.

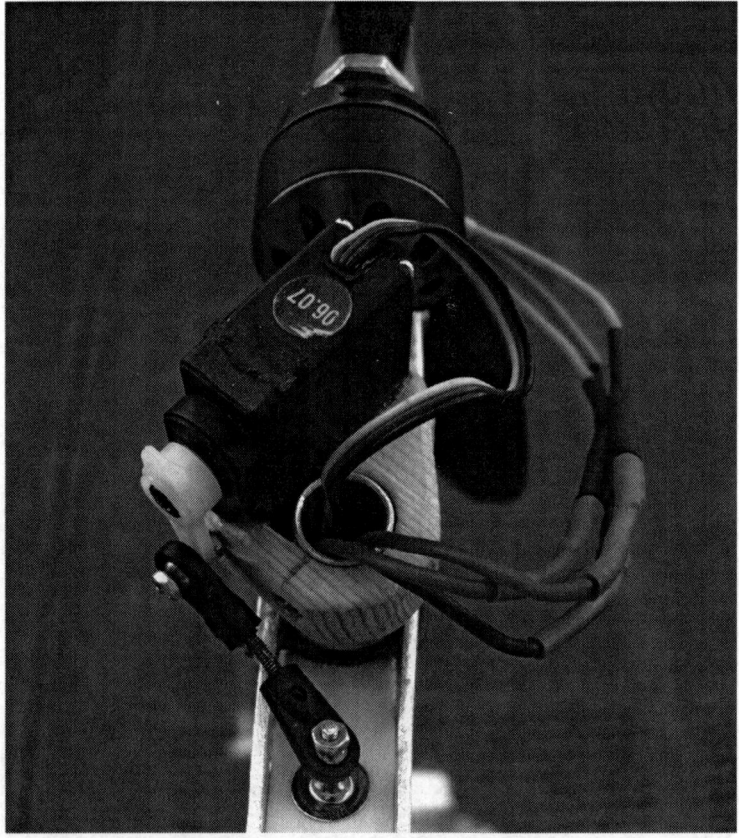

Figure 43: Pivoting rigger, Photo: Karl-Heinz Wolf

The consequence is that a drive must be pivotally designed. In this way, a compensation of yaw torque or, if desired, a rotation around the yaw axis can be achieved. The Tricopter in Figures 42 and 43 was built by Karl-Heinz Wolf.

The control of the angle φ in the nick and roll axis is a bit more complicated. The two axes are no longer decoupled. Imagine that the pivoting rigger of the model is 'rear'. Then the other two riggers are 'front' (actually, front-right and front-left). In order to steer the nick axis, both front drives need to act together, whereas in the rear there is only the 'rear' drive. To control the roll axis, the drives 'front-left' and 'front-right' operate.

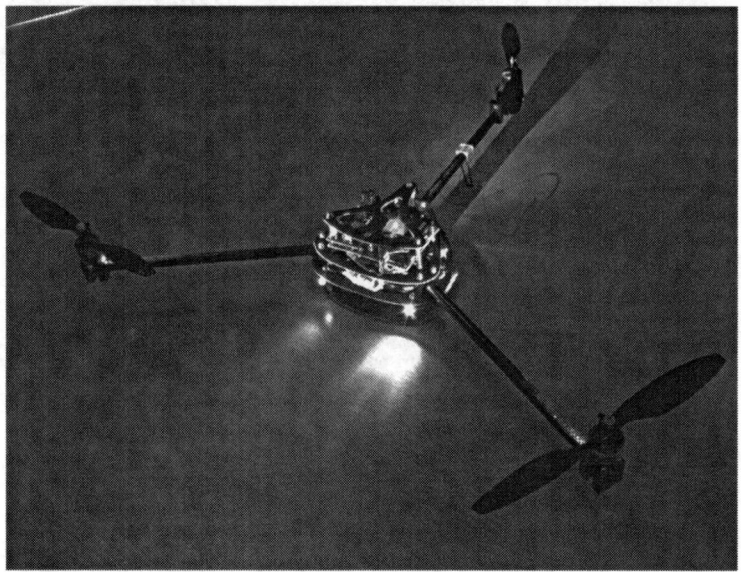

Figure 44: Tricopter with carbon riggers, Photo: William Thielicke

The two front riggers therefore have a double function: They must simultaneously stabilize the nick and the roll axis. To control the nick axis they both produce at the same time more or less thrust, whilst to control the roll axis one drive produces more thrust and the other less. Of course, with a programmable electronic control system this can be easily implemented.

Figure 45: Pivoting rigger, Photo: William Thielicke

Figures 44 and 45 show a further Tricopter with a pivoting rigger from William Thielicke. The frame and the rigger were realized very beautifully and shaped with carbon.

Advantages of Tricopters:
- They need only three motors, propellers, riggers and brushless controllers, which means a corresponding reduction in price.
- You can use right-turning propellers, of which there is a much greater choice in the market.

Disadvantages of Tricopters:
- One axis must provide pivoting.
- The control is more complicated because the nick and roll axes are no longer decoupled.

Y6 configuration

The system shown in Figure 46 is actually also a Tricopter. At least it has three riggers. The difference to the Tricopters presented above is that not only three but six motors and propellers are mounted. 'Y' stands for 'three riggers' and '6' stands for 'six motors and propellers'. As the picture shows, a left- and right- rotating propeller is always mounted to the bottom and the top.

The control in the nick and roll axis is the same as explained above. The two rotors of an axis both rotate at the same speed. In contrast to above a pivoting axis is now no longer required. For yaw, one of the two propellers of an axis simply turns faster and the other slower. The downward propellers, however, are a little more vulnerable to breaking on landing. In addition, the price advantage of the above-described Tricopter here no longer applies. Even more motors and propellers are needed than with a Quadrocopter.

Nevertheless, the Y6 configuration is a very common special form. The system shown in the photo is from Michael Sachs (power-frame.de). Another special feature is the propeller made of carbon. These are good to use because they are very strong and particularly light. This also leads to a small mass moment of inertia.

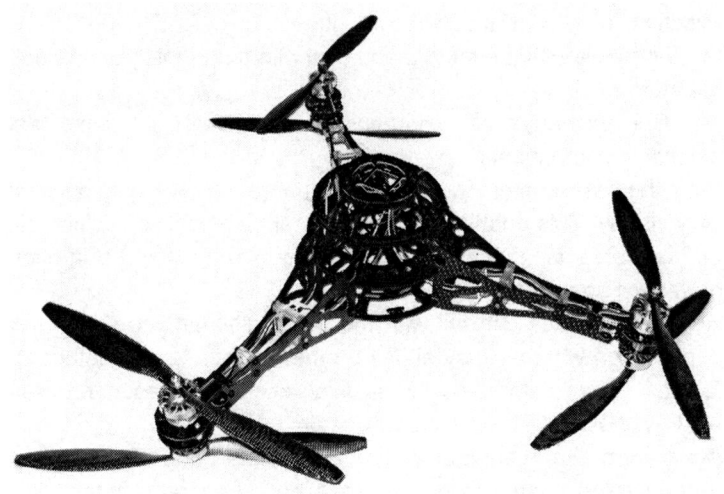

Figure 46: Y6 configuration, Photo: Michael Sachs

7.2 Hexacopters, Octocopters

Many technical systems are designed as redundant for safety reasons. This means that individual components can fail without affecting the overall function. Hexa- and Octocopters are an excellent example of this.

If during the flight with a Quadrocopter a brushless motor or controller fails, it unavoidably sags the rigger, and a rollover is the fatal consequence. Auto rotation, which is often seen in emergency cases with helicopters, is not possible here. The small moment of inertia of the propellers vitally important for controlling is in this case just a disadvantage.

These failures are very rare and when they happen it is almost always due to a wrong combination of motor and propeller. A camera pilot is often hovering around with a heavy and expensive SLR camera. It is understandable that he fears this terrifying scenario. For this reason, Hexacopters with six or Octocopters with eight propellers are built.

Figures 47 and 48 show the Hexacopter and Octocopter of Peter Plischka. They consist of two components:

 a) The frame with the motors and controllers (mounted underneath the motors)

 b) The exchangeable electronics dome, with all base and extension components

With the Hexacopter, the bottom-mounted camera can be seen very clearly. This enables images from all perspectives, which can be generated without interfering propellers. The landing skid offers protection against damage.

How many drives can fail with the model still landing safely also depends on whether they all fail on the same side or whether the failures occur symmetrically. In any event, a reasonably safe landing should still be possible if one drive fails.

Apart from the redundancy, the Hexacopter and Octocopter – assuming the same motors and propellers – generates in total 6 / 4 = 1.5 and 8 / 4 = 2 times more thrust than the Quadrocopter.

Figure 47: Hexacopter with camera, Photo: Johannes Thor

Figure 48: Octocopter, Photo: Melanie Plischka

Advantages of Hexacopters and Octocopters:
- Due to the redundant design of the motors and propellers, at least one drive can usually fail in flight without the copter suffering damage.
- With the same motors and propellers more thrust will be generated in total than with a Quadrocopter.

Disadvantages of the Hexacopters and Octocopters:
- The greater number of motors, propellers, riggers and brushless controls corresponds to a higher price.
- As more components are needed, the construction, putting into service and maintenance is more complex.

7.3 Depron bodies

Figure 49: Larger models look almost like real, Photo: Heinz Büchi

So that the orientation in the air is more visible, the models can be fitted with bodies. With model aircraft, the fuselage is responsible for the correct fit of the wings; that means 'form follows function'.
This does not apply here. The bodies – made of Depron, painted paper or stringed carbon – have no special function. They are actually only ballast. But exactly this may be interesting. Due to this

fact, the imagination knows no limits. The model of an 'air cabriolet' (in reality, for now, non-existing) shown in Figure 49 consists of 3mm Depron. To allow a good assembly, the middle of the interior was strengthened with some balsa strips.

In speed flight there is still a bit of 'function' felt. At higher speeds, the rigid rudder helps a little to keep the forward direction. The weight is 75g in total, including a model pilot. In the case of an 800g copter this is barely noticeable. Basically a Quadrocopter with a thrust–weight ratio of 3:1 can take a little more than half its own weight. Nevertheless, every gram in the Depron body should be paid attention to.

Figure 50 shows the 'United Starship Enterprise' from the 1960s television series 'Star Trek'. The entire structure is about 300g, 120cm long and 50cm wide. The 'warp gondolas' which can be seen in the back are made of painted and rolled paper, while the front part is made of 3mm painted Depron with a diameter of 50cm. The whole structure is held together with carbon rods. It is a little susceptible to wind gusts, but was already successfully used at model helicopter flight events with moderate wind.

The models in Figures 51 and 52 are smaller, but also made from 3mm Depron. The 'Galileo' and the 'Work bee' are also models of small space ships from 'Star Trek'. With just 25g weight, they also fit Quadrocopters with 30s frames and 30g motors.

The rocket in Figure 53 is made of painted paper. The total weight is only 25g. The pilot body is also Depron. When the Quadrocopter is attached for forward flight, the rocket flies almost in horizontal position. That looks pretty good. Due to the aerodynamic shape, it is not susceptible to wind. Also this construction can easily be used with 30s frames and 30g motors.

Figure 50: Semi-scale? The U.S.S. (United Starship) Enterprise will in any event be launched in the year 2245. 'Who owns one, doesn't need to believe', Photo: Heinz Büchi

Figure 51: Galileo, the shuttle of U.S.S. Enterprise, Photo: Heinz Büchi

Figure 52: The 'Work bee' is used for the construction and repair of star ships.

Figure 53: The rocket is made of painted paper; the weight of the structure is 25g

Figure 54: UFO in European airspace? Photo: Heinz Büchi

The UFO shown in Figure 54 has a frame made of 2mm-thick carbon rings. They are held together with balsa strips. Painted paper gives the model the unusual appearance. The weight is 75g with a diameter of 1m. A peculiarity with this construction is that the Quadrocopter with a 30s frame and 30g motors disappears completely. This is only possible because it is open above and below and therefore the propeller airflow is not disrupted.

7.4   Aerial Sedan

The helicopter pioneers Stanley Hiller and Igor Sikorsky knew that helicopters, from the energetic point of view, would never match airplanes. They would need more power to keep themselves in the air. Nevertheless, both developed these aircraft up to today's perfection.
They also knew that with helicopters you can solve tasks that are impossible to accomplish with airplanes. A helicopter can stop at

any point in the air. Therefore, it has become an indispensable tool in many fields.

In the 1950s also the vision of individual traffic with a four- rotor helicopter, an early Quadrocopter, emerged.

The vision

In 1957, the July issue of the magazine 'Popular Mechanics' contained an article about 'Hiller's Aerial Sedan, Your Flying Car for 1967'. It was announced that almost everybody could buy such an aircraft in 10 years. The functioning was clearly explained, namely that for forward flight the speed-controlled front propellers would rotate for a short time slightly slower than the rear. Braking would be achieved by a slower rotation of the rear propeller. There were at that time some four-rotor prototypes for the army, such as a flying jeep or the X-19 by Curtiss-Wright.

If one wanted to go into this market, then it was clear to everyone that these VTOLs (Vertical Take Offs and Landings) would have to be improved also constructively.

Quadrocopters with their vertically arranged rotors represent a considerable safety risk. One must only remember that passengers would have to get in and get out between turning rotors. Furthermore, in the case of private traffic, it must be possible to start and to land safely everywhere, even near to people. The solution to this is to integrate all rotors completely into the bodywork.

The disillusion

Of course, they brought the flying jeeps into the air as well as the X-19. But they soon had to recognize that the technology was just not mature enough to build something like an air car for everyone. It was also missing the microprocessor technology, which is standard in today's Quadrocopter models. Together with accurate and drift-free sensors and powerful motors, they allow robust stabilization for the first time.

Therefore, in those days only experienced pilots could steer these prototypes. Various setbacks and crashes finally led to the fact that this vision remains unfulfilled to this day. The Aerial Sedan, the flying car for 1967, was never built. To date, individual transport takes place largely on wheels and on the ground. The third dimension is, as already at that time, reserved only for a few.

The model

Nevertheless, an Aerial Sedan can be built as a model. Figure 55 shows this in 'semi-scale'.

Figure 55: Aerial Sedan as 'semi-scale' model

As a basic platform, commercially available electronics are used. The motors and propellers are Hacker A20-22-L and 10" x 4.7". This combination generates with all four drives together a little more than 2kg (20N) of thrust. This is enough to build a nice model with a total weight of about 1kg, and to maneuver it safely in the air.

The basic structure of carbon fiber tubes

The basic structure is realized with carbon fiber tubes, the same as described in Section 2.8. The control board, the receiver and the four brushless controllers are all placed on the center plate. This is made out of the sandwich with two 1mm carbon fiber panels and an aluminum cross of two 10mm square tubes. This can, on the one hand, take the 8mm carbon tubes of the riggers and ensure, on the other hand, that the electronics will not be damaged in a crash.

Figure 56: Carbon construction and safety rings for the 'Aerial Sedan'

Below are four 6mm nylon screws, which provide for a sufficient distance from the ground. The Quadrotor stands on the wing bolts on the bottom. Here is also enough space available to include a 3S-2100mAh-LiPo accumulator. The electronics are also protected from the top, using nylon bolts and pieces of carbon fiber. Since the length of the 'Aerial Sedan's' propellers are larger than their width, two longitudinal tubes are attached. At their ends they take the drives. Two crossbars complete the construction. They are responsible for ensuring that, in operation, the shafts of the motors do not twist against each other. Figure 56 shows the construction.

Safety

What is true for a (never built) original, applies also for the model. When the company Silverlit brought the X-UFO to the market in 2005, the relatively soft 8" propellers were still protected with EPP rings. Today Quadrocopters are often operated with larger motors and 10" propellers. These are also sometimes relatively heavy. Flying Quadrocopters, however, must not be a safety risk for spectators and pilots. That is reason enough to treat this issue deeper. In the present model, rings of 50mm x 930mm x 0.5mm carbon fiber are mounted as a propeller guard. The extra weight of all the rings is 120g, which makes up about 10% of the total weight of the model. It would be nice if this 120g was used in each of today's flying Quadrocopters for safety!

Figure 57: Propeller core

The body made of Depron

Unfortunately, the finish of painted Depron often looks a little like bricolage. Therefore, a new method with the use of a self-adhesive foil is followed. Its adhesion to the Depron is not quite as good as if it were glued to wood. If, however, the foil is sitting right the first time and is not removed for correction purposes, then it glues quite well and gives the models the same look as if it were disguising any other material. With this combination, a very light and elegant body can be built. Unlike with most other flying models, it has no supporting function. It is just the carbon tube construction which provides for that.

The hood is built more conventionally from 5mm x 5mm balsa strips. The roof consists of covering paper, which is finished with red spray paint.

All in all, Quadrocopters would be much better accepted among model-makers if they were a little more colorful. Many pilots say, however, that all the work building a Depron body would be destroyed in a rollover. However, anyone who flies Quadrocopters today notices that the technology has advanced so far and so safely that (semi-)scale models can be built without problems in this regard. The only problem that arises is that there are not as many flying originals.

In addition, better visibility and orientation is also ensured with such a model. Many pilots can only fly their Quadrocopter a few meters away from themselves, because otherwise they can no longer see where the front of their model is. A Depron construction gives a completely new feeling of flight, and flight radiuses are also possible which certainly come close to those of normal model airplanes.

7.5 Private transport with Quadrocopters from today's perspective

One may wonder whether it would be possible from today's perspective to extend the private transport in the third dimension using Quadrocopters. Basically the idea is still great. It would change the world at least twice as much as the invention of the automobile itself.

There is technically no reason today why the 'Aerial Sedan' should not be built for real, for use as private transportation. The vision would be technically feasible. The controller could be designed with assistance sensors so that anyone with a driver's license would be able to maneuver it safely through the air.

However, there are other aspects that must be considered. The dimensions of a modern automobile would be the graduated measuring rod by the construction. With a width of about 2m, the rotors could therefore have a diameter of 1m each. With a lightweight construction, a take-off weight of 500kg may be achieved. Applying a little theory, one finds that more than 100kW would be needed just to float. Since a Quadrocopter still requires reserve power for the regulation and forward flight, a 250kW motor would be sufficient.

As discussed in Chapter 6, the specific thrust increases with a larger rotor area. Therefore, the rotors of helicopters are indeed very large. So if one could assume a greater rotor diameter than 1m, the power balance would be much better. However, as in the 21st century energy is increasingly developing into a resource problem, it appears to make little sense to use the Quadrocopter for private transport.

Another question would be to clarify what would happen if the drives fail. A helicopter can land safely under these circumstances with autorotation. A Quadrocopter, which owns speed-controlled rotors, has propellers with small inertia. The regulators must be able to change the speed quickly in order to provide a stable flight. Autorotation is therefore not possible. So if a rotor fails, a rollover is the fatal consequence.

A car on four wheels doesn't have the above problems. If the drive fails, it simply doesn't move anymore. The energy balance is much better, since no power is needed for hovering. All energy can be spent on forward movement.

In summary, it must be said inform the current perspective: The Quadrocopter is for the time being reserved for model flying and unmanned flying. There, its potential is very large. From an energetic point of view and because of the safety considerations in the event of the failure of the rotors, it does not make sense for use as private transportation.

8. Initial operation, sources of error and first flight

After all the cables are soldered according to the instructions and all motors are bolted properly, comes the great moment of the first start. Here is the most important rule:

The initial operation should be made without propellers for safety reasons. Once the propellers are mounted, guard rings should be used.

Since the settings have not been made, we do not know what software is on the system or whether the sticks of the remote control must now be inverted or not (especially the gas lever!). So we should be careful. The motor after all powers to a total of at least a few hundred watts. When these forces are uncontrolled, broken propeller blades and bent motor shafts are still the most harmless imaginable damages.
Even worse is the assumption that one could easily hold a Quadrocopter in one's hand, and then turn it on. Even the smaller systems with 30g motors can easily break free from hand grip if something is misconfigured.
If the propellers are not yet attached, it can be safely turned on. The configuration by PC can now be done through the interface. Instructions are available from different manufacturers. These can usually be downloaded from the Internet.

8.1 Check the functions on the ground

If everything was done right, the motors can be turned on by a button or stick movement. They run fine in idle when the throttle stick of the remote control is down. Still without propellers, the gas functions can now be tested. With a higher throttle position all the four motors should turn faster in equal measure. A running-in of the brushless motors is not necessary. With direct current motors, emptying one or two battery charges at half throttle had formerly increased the lifetime. In this way, the brushes were shaped correctly. Without propellers, the throttle stick should not be moved to full throttle, because the speeds would become too high and the bearings might become hot.

Now the nick and the roll functions can be tested. Once the stick is pushed in one direction, this motor should rotate more slowly, and in the opposite direction faster. With the yaw functions two opposing motors must always rotate faster or slower.

The 'dry run' without propellers seems a bit boring and unspectacular. Anyone who has performed the initial operation already with propellers (and there are many, because one can't wait until the Quadrocopter flies) confirms, however, that he would have avoided many broken propellers, defects of one or the other part or even wounds had he performed the tests without the propellers attached.

If something does not work you should check the following using the PC interface:

- *Is what you think is the 'front' really the Quadrocopter's front?*
- *Are all the remote control sticks assigned to the correct functions?*
- *Are all remote control sticks configured in the right direction (especially gas)?*
- *Are the brushless controllers properly assigned according to the riggers?*

8.2 Range test of the remote control

With Quadrocopters the electrically conductive carbon is used as a framework. Therefore, the range is usually somewhat restricted compared with model airplanes. For the optimal function, the receiver antenna doesn't like having any other conductive material around.

As a range test, many propose that with a retracted (not unscrewed) antenna a distance of 50m is possible without interference. Disturbances usually first show up when the motors are started. Then current peaks occur, because the brushless controller chops the DC voltage in order to produce sinusoidal signals. These peaks produce electric and magnetic fields at all frequencies, including at the wavelength of the incoming signal of the remote control. If a helper is present, he should move with the model and provide feedback. In this way the behavior with started motors can also be tested.

Here again, a system configurable by PC shows benefits. With some systems, the quality of the received signal can be tracked and it can then be read via the interface – similar to a flight recorder.

Bidirectional remote controls with a 2.4GHz base often have functions that allow the quality of the received signal to be read directly on the remote control. This is another plus point for this technology.

In the event of an error, the antenna of the receiver is often responsible

- *Is it mounted unrolled?*
- *Has it possibly suffered a broken wire (test it using a multi-meter)?*

8.3 Mounting the propellers

Once all these tests have been carried out successfully, the assembly of the propellers follows. Here is another possible source of error. Strictly two right- and left-turning propellers must be installed. They must also be mounted in the right place and the same rotating ones must always be opposite to each other.

Next there is the source of error in rotation. The drives have to rotate in such a way that both the left- and the right-turning propellers generate thrust downwards.

The rotation can be reversed with most systems via the PC interface, or any two of the three motor wires can be interchanged in order to achieve the same effect.

When everything is done correctly, the Quadrocopter should now fly as desired.

In the event of an error, check the following:

- *Are the two left- and right-rotating propellers mounted in the right place?*
- *Do they turn in the right direction?*

Vibrations

It may happen that vibrations occur during flight. One or both axes then jitter. This topic is also discussed again and again in relevant internet forums. The reasons for this can be control electronics, control parameters or sensors. The cause can often be found in the motor/propeller combination or their assembly. The drive requires a high acceleration, for high dynamics, so that the axes can be controlled quickly enough.

Many propellers warp during this acceleration phase and therefore distort the regulation. Examples are the propellers of coaxial helicopters. In principle, they would be an ideal choice because right- and left-turning propellers are always available. However, there are very few Quadrocopters which fly reliably using such propellers.

Further, the propeller imbalance can cause problems. It is hard to believe that many propellers from over the shop counter have a horrible imbalance. A cheap propeller balance and a little sandpaper can sometimes really work wonders. Alternatively the propeller can be mounted on a spare shaft. This is placed on the two shoes of a vise, whilst the propeller sits in between. The end of the propeller is always sanded. This has the greatest influence on the unbalance.

8.4 First flight

During the first flight there should be plenty of space around you and no wind if possible. Ideal would be a huge sports hall. As this is generally not available, a courtyard or a parking area in the evening, when there is almost no wind, is also sufficient.
The configuration '+' is best to start, because then any rigger corresponds precisely to one of the directions 'forward', 'backward', 'left' and 'right'. The front arm should point away from you. For a better identification, you can mark it with a piece of colored tape.
Since the gyro sensors based on piezo technology are temperature dependent, a calibration should first be performed. Here, the signal for angular velocity of zero is first measured and then stored. It will serve afterwards as a reference. Some control electronics perform the calibration independently, while for others it can be enforced with a certain movement of the remote control stick.
Once the motors are started, you should check in each case once again the correct control functions on the ground. If you push the nick and roll stick in a certain direction, the corresponding rigger should tilt there. If you give more gas, all propellers should rotate faster.
Now it can be started. With a short gas kick, some height should immediately be gained. This sounds strange at first, but near the ground the ground effect is noticeable and the Quadrocopter gets into its own airflow. Meaningful practice is guaranteed only from a safe distance of about one meter in height.
In everything that follows, this is the most important thing to remember: If you lose control, reduce the gas immediately. A Quadrocopter which drops from one meter onto the ground seldom suffers major damage.

Yaw

First you should try to keep your model in a floating position. It will initially not even be important that it drifts off to the right or left, because you have sufficient space around you. Rather, you should concentrate first of all on the yaw stick and move it so that the 'nose', the front rigger, is always pointing away from you. This is the only way to ensure that all stick movements of the remote control move the model in the desired direction. Imagine that the nose points in your direction. Then all the movements of the remote control stick would precisely steer the model in the opposite direction. This is only useful for experienced model pilots and not for a beginner. Now you can play with the trim. If the stick position is in the middle, the nose should turn as little as possible. Since the Quadrocopter is constructed symmetrically, it should not be necessary to trim strongly.

Hovering

After having practiced the yaw and, after having had several times to get your model in the correct orientation but far away from you, you should next try, while operating with the yaw stick, also to move the nick and roll stick so that finally the model hovers without drifting away to the side, forward or backward. If appropriate, it will also be trimmed a little.
It takes a few battery charges to master it. Afterwards you are still not a perfect pilot, but on your way.
After the Quadrocopter hovers on the spot and, thanks to the correct yaw, always moves the riggers parallel to the remote control sticks, one can now try to yaw a little away, and rotate together with the model. This will be practiced as long as necessary to get used to rotating together with the model around its own axis without getting dizzy.
Next is yaw without turning yourself. Once again, we begin first with small angles and try to keep the model on the place of hovering. The difficulty now lies in the fact that the riggers no

longer move parallel with the remote control sticks. Anyone who has driven a remote-controlled model car can attest to that. The difference to a Quadrocopter, however, is that there are only two channels (gas and steering wheel), whereas here four of them (gas, yaw, nick and roll) need to be controlled simultaneously. This is a much more difficult task. Later, the angle will be increased gradually until one masters the so-called 'nose hovering'. Here, the nose points in your direction and all functions are reversed.

Then almost all the possibilities are open, such as the flying of a circle or a figure of eight, etc.

Not all Quadrocopter pilots master this. It's like with everything. You learn it at a young age a lot more easily than later. Many of the pilots are just hovering without yawing a lot. The important thing is that everybody has fun with his hobby, and mounts cameras or compasses or even installs GPS on the systems and enjoys the wonderful technology that makes all this possible.

9. Literature

(1) Der Standschub von Propellern und Rotoren, Helmut Schenk, 2002

(2) Integrierte Navigationssysteme. Sensordatenfusion, GPS und Inertiale Navigation, Jan Wendel, Verlag Oldenbourg 2007, ISBN: 978-3-486-58160-7

(3) Global Positioning Systems, Inertial Navigation, and Integration, Mohinder S. Grewal, Lawrence R. Weill, Angus P. Andrews, Verlag: John Wiley & Sons, Inc. 2007, ISBN: 978-0-470-04190-1

(4) Das Depron Buch, Hinrik Schulte, Verlag für Technik und Handwerk GmbH 2008, ISBN 978-3-88180-741-8

(5) Moderne Fernsteuerungen für RC-Flugmodelle, Manfred-Dieter Kotting, Verlag für Technik und Handwerk GmbH, ISBN: 978-3-88180-780-7

(6) Das Lipo-Buch, Ulrich Passern, Verlag für Technik und Handwerk GmbH, ISBN: 978-3-88180-781-4

(7) Selbstbau von Brushless-Aussenläufer-Motoren, Heinrich Hilgers, Neckar-Verlag 2006, ISBN 978-3-78830-683-0

(8) Regelungstechnik, Einführung in die Methoden und ihre Anwendung, Otto Föllinger, Hüthig Verlag 2008, ISBN: 978-3-7785-2970-6

(9) AVR: Hardware und C-Programmierung in der Praxis, Florian Schäffer, Elektor-Verlag 2008, ISBN 978-3-8957-6200-0

Lightning Source UK Ltd.
Milton Keynes UK
UKOW03f2113161213

223137UK00009B/157/P